ATLANTIC REVERBERATIONS

To my parents, Darryl and Kathryn Adams
for fostering an interest in what lies beyond the horizon

Atlantic Reverberations
French Representations of an American Presidential Election

PAUL C. ADAMS
University of Texas at Austin, USA

Routledge
Taylor & Francis Group
LONDON AND NEW YORK

First published 2007 by Ashgate Publishing

Reissued 2018 by Routledge
2 Park Square, Milton Park, Abingdon, Oxon OX14 4RN 711
Third Avenue, New York, NY 10017, USA

Routledge is an imprint of the Taylor & Francis Group, an informa business

First issued in paperback 2018

A Library of Congress record exists under LC control number: 2006032978

ISBN 13: 978-0-8153-8767-1 (hbk)
ISBN 13: 978-1-1386-1890-9 (pbk)
ISBN 13: 978-1-3511-6288-3 (ebk)

Contents

List of Figures

List of Tables

Preface

One French observer of the 2004 American presidential election framed the outcome in the following way: "The United States has missed its rendezvous with history. In my view, this day [November 2, 2004] will remain, no less than September 11, 2001, an indication of the start of the 21st century, and an equivalent catastrophe." (Climacus 2004). The distinctions between Us and Them that are so vital to concepts of nation and nationality are brought into public discourse from time to time by certain critical events. These events – whether violent like revolutions or peaceful like elections – are turning points for the construction of national identities both in and beyond the states directly concerned. Dijkink (1998) calls these salient events "peak experiences."

The attacks of September 11 constituted such a peak experience in the US and abroad as did the 2004 election to which "Climacus" (an Internet pseudonym) linked them. The importance of these events spilled across national boundaries to virtually all the rest of the world, prompting varied debates. In Europe, the questions raised by the attacks were particularly pressing: was the US part of "The West" (in France *l'occident*) or was the US a separate entity, pursuing its own political and cultural path, useful to Europe but also dangerous – no longer wedded to Europe through the ties that had existed at one time?

This geopolitical debate intensified with the 2004 presidential campaign. What would be the implications of a second term for George W. Bush or an upset by John Kerry? Why was half of the American population supportive of Bush and why weren't more people in favor of his opponent who was widely favored in Europe? What did the closeness of the election say about America? Was the Bush administration a threat to global stability, and in what way? Finally, what should or could the French do in response to the political climate in the last remaining superpower, what the French statesman Hubert Védrine called the *l'hyperpuissance* (the hyperpower)?

Americans need to engage seriously with such foreign perceptions, on the one hand, to address pragmatic US security interests and, on the other hand, to more effectively support the ideals that the US has championed, albeit with less than perfect integrity, for many years, including freedom and democracy. The most cynical observers see these ideals as no better than a smokescreen for American imperial ambitions, but many others see the ideals functioning ambivalently. At times they mask the operation of US power but at other times they provide a moral order that inspires Americans to act in unselfish ways. There is also, of course, the view that America can do no wrong, that "we are the liberators, we protect the weak and preserve the innocent" (Colombani and Wells 2004, 7) a view that Americans have absorbed through indoctrination until foreign scholars ironically refer to it as part of their "DNA code." But scholars at least must vigorously question the motivations behind the superhero complex that drives Americans to assert their vision of order on the world, and drives foreign observers to produce critical responses.

Toward this end, I examine discursive constructions of the American presidential election of 2004 by French observers in three distinct communication contexts: television, newspapers, and the Internet. While all three media indicate that the French preferred John Kerry over George Bush, the cross-media analysis reveals many subtle gradations in the preference, nuances in its implications, and diversity in its expression. I argue that these nuances reflect intellectual debates relating to the ideas of Europe as a unified world power and the US as a uniquely unstoppable global power. In addition, I show that as the political drama reverberated across the Atlantic Ocean through a set of rather different technologies, each demonstrated its own peculiar way of mobilizing representations of the US for and by its various audiences.

The book is designed to encourage both theoretical and pragmatic discussions about the future of the geopolitical system – about the web of power relations between 21st century states. It is based on the premise that American "leadership" of the world through force and economic coercion cannot be considered democratic simply because the US is internally a democratic state. Voices expressing concern with US hegemony, like many of the French observers considered here, are not opposing democracy; rather they are expanding the democratic process to a trans-border space of debate and discourse – participating in the on-going evolution of citizenship in the face of globalization.

I want to thank the following individuals who helped make this project possible. The University of Texas provided a "Maymester" teaching assignment in France in 2004 and funded two subsequent visits partially or wholly. The University funding included a Summer Research Assignment by the Office of Graduate Studies. Mme. Blandine Perrin of API, Grenoble generously committed her time and office space to many, many copies of *Le Monde* in fall of 2004, as well as devoting considerable energy to mailing clippings from these newspapers to me in regular installments. Mme. Danielle Pheng-Gaboriaud demonstrated equal generosity by collecting and forwarding articles from the odd couple, *Libération* and *Le Figaro*. Danielle again deserves thanks as does Mme. Geeta Rocard for providing comfortable and affordable accommodations in Paris, and for taking a remarkable degree of care to make sure I had a working knowledge of the neighborhoods in which I was living. Thanks also are due to Mme. Divina Frau-Meigs who went out of her way to introduce me to the remarkable collection of audio-visual materials at Inathèque, without which the television portion of this research would not have been possible. The research assistants at Inathèque were also very helpful and tolerant of my requests. Finally Kaitlin and Karina have my boundless gratitude for putting up with a project that postponed our next trip to Quebec, put me in a foul mood more than once, and used up many hard-earned frequent flyer miles.

Paul C. Adams
Austin, TX, August 2006

Chapter 1

The International Echo Chamber

Atlantic reverberation is my term for the way geopolitical discourses echo across the Atlantic Ocean between the US and Europe, sounding and resounding. Consider an example: in January of 2003 the French and Germans reject US arguments about weapons of mass destruction in Iraq, leading US Defense Secretary Donald Rumsfeld to brand them as "old Europe," which in turn generates a flurry of annoyance in European media. Americans, too, are annoyed and over the following 22 months the French and John Kerry are both mocked in digitally-edited images that "Frenchify" John Kerry. The mocking images are in turn discovered by a French "blogger" whose website makes them accessible to thousands of Internet users who lie outside the images' intended "audience." The digital images precipitate comments about the prejudices of Americans, their xenophobia and their French-bashing, including one comment written in English, most likely by a Francophile American "lurking" on the website: "Bush will lose in november [sic], rest assured." Reverberation can be heard in the form of echoes between countries and between official government statements and grassroots communications.

Meanwhile, messages also reverberated in the somewhat more ethereal circuits between American and French intellectuals. French political theorist Étienne Balibar counted four notable scholars based at least partially in the United States who were calling on Europe to mount concerted opposition to the excessive power of the US (Balibar 2005, 17–24). The author of *L'Europe, l'Amérique, la Guerre* (Europe, America, the war) used these authors to give a special kind of legitimacy to his argument that Europe should provide an alternative to the American model of democracy. Appropriating American discourses in a different way, French Marxist Alain Joxe critiqued the writings of American political and social theorists Alvin Toffler, Samuel Huntington, and Anthony Lake (2004a, 107–118). In addition, as George W. Bush was commencing his second term in office, scholarly French periodicals such as *Le Débat* and *Commentaire* sought to elucidate this turn of events by reprinting translations of the writings of various American political theorists.

Although we are accustomed to thinking of political processes as occurring in containers, such as countries or states, the ground rules of politics are changing. Media of communication such as books, newspapers, television and the Internet host an on-going political discussion that is not just international but intercontinental and intercultural. Thus *reverberation,* constituted through the paired acts of writing and reading across languages, continents and political cultures is a fundamental component of democracy. In what has been referred to as a "deterritorializing world order" (O'Tuathail and Luke 1994, 383) people often discuss territories – political containers like France and the United States – but their discussion crosses national borders and exists in the virtual space of communication media (Adams 2004). Ironically, the spaces in which we talk are much less bounded than the spaces we talk about.

Discourses of French observers about the *hyperpuissance* (hyperpower), American political speeches appropriated by French observers and then discussed, and the subsequent reverberation of meanings across the Atlantic – these are the basic elements of the story of 2004 that I wish to examine. Generally speaking, the book is concerned with the representation of a powerful country by observers in a less powerful country, which is a perspective that, for whatever reason, remains rare within the promising field of critical geopolitics.

The diffusion of electronic communication technologies ranging from radio to the Internet has supplemented the virtual places created by books and their audiences, helping (along with international capital flows, migration, and trade) to make nations very leaky containers. Geopolitical discourses now reverberate in a multitude of communication spaces stretched between nations via a wide range of media, exacerbating tensions between national populations with different interests and worldviews but also forming ephemeral supranational communities that dissipate tension. Particular reverberations like the discussion of an American presidential election may emerge and subside in a matter of weeks or months. Nonetheless they form threads in a tapestry of trans-border communications, marking the first steps toward political coordination of a post-state world and the globalization of the public sphere (Kellner 2000; Calhoun 2003). The creation of a transborder political discussion can have significant effects on subsequent political mobilizations within bounded political containers (Adams 1996). Its existence contributes not only to the parallel evolution of what Peter Taylor calls "internationality," "interstateness," and "interterritoriality," (1995) but also to the formation of what Jürgen Habermas (1991) calls the public sphere. And it does so, strikingly, at ever-expanding scales.

This international echo chamber demands careful analysis and coordinated response, regardless of one's political position. It is one reason why American attitudes of belligerence and arrogance, and the rhetoric of "good" and "evil," "with us" and "against us" deployed since 9–11–2001, have been serious security liabilities. Such harsh words resound like firecrackers in an echo chamber. Their echo is heard, for example, when French observers brand the US with the label of "empire" (Joxe 2004a; Wieviorka 2004). An understanding of the feedback loops that rattle international communication suggests that more than ever the US must follow Theodore Roosevelt's advice and speak softly, even if it carries the biggest stick. But perhaps that is not enough, as the size of the stick is motivating a groundswell of international debate about "democratic deficit."

To take a larger perspective on this situation, geopolitical debates such as used to take place in the conference rooms of diplomats now occupy a space between places, a virtual space that transcends the familiar geography of bounded places. This virtual space of political discourse is relatively easy for "ordinary" people in developed countries to access, and although such communications are dismissed by many

Americans as meaningless chatter (especially when it includes critique of American policy) deterritorialized discussions and distanciated political forums are resources for building a new architecture of international consensus (O'Tuathail and Luke 1994; Giddens 1984; Habermas 1999). Whether we prefer to call them "ideoscapes" (Appadurai 1996) or "flowmations" (Luke and O'Tuathail 1998) structured flows of international communication are increasingly important to political and ideological debates throughout the world, and while the geopolitical imagination works to defend the sovereignty of the nation-state against deterritorializing processes – flows of money, people, goods and ideas – by which the nation-state becomes "hollowed out," the same imagination is also at work questioning particular states (particularly the United States) and their power over the international or interstate system (Taylor 1995; O'Tuathail and Dalby 1998).

The howl of feedback in a microphone is reverberation without a human link, but the reverberation that forms the focus of this book depends on human links. These links are not just people but citizens, which means *people with certain expectations and beliefs about their role vis-à-vis society*. Citizens share a desire to hear and be heard. The kinds of discordant reverberation described above are not just "technical difficulties" that need to be *solved* by top-down control or propaganda, governmental or otherwise. On the contrary, they are components of the democratic process that are overreaching the bounds of formal democratic institutions, suggesting faltering steps toward "democracy unbound" (Low 1997). As Jürgen Habermas argues in his many works, the search for mutual understanding is an unending process that lies at the heart of civil society, that is based on the premise of equal access to communication, and that continually redraws the boundaries in which people understand their world (Habermas 1999; 1991; Adams 2005, 130–141). We must therefore begin thinking of democracy without thinking automatically of territorial boundaries, however great the theoretical difficulty in doing so, because politics continually re-defines boundaries rather than being neatly contained within boundaries (Jonas 1994; 2004).

I believe it will be extremely difficult for American leaders to take into account public opinion around the world, practically as well as politically. But as democratic discussion and debate inexorably slip the bounds of the nation state and take up residence in larger circuits of interaction and coordination, listening becomes a prerequisite of national security for Americans as much as for others.

In this introductory chapter we prepare for an excursion into French views of "America" by considering the ways in which citizenship is evolving as a result of greater interconnection and globalization and we reflect on the role played by various media in this re-bounding of citizenship. First it is necessary to comment on the choice of French discourses as our "lens" on America and on worldviews of America.

France is one of the strongest European countries – forming with the UK and Germany a "triangle" of leading European powers. With the fifth largest economy in the world and one of the strongest post-colonial networks of cultural and political alliance, as well as overseas departments and territories in the Caribbean, Indian

Ocean, North America and South Pacific, France contributes a major portion of Europe's international economic, political and military stature. France's political legacy is enriched by the artistic, literary, scientific and philosophical contributions of the French people, including the foundations of modernism (Comte, Descartes, Voltaire, Picasso), romanticism (Rousseau), existentialism (Camus, Sartre), feminism (Beauvoir) and postmodernism (Baudrillard, Bourdieu, Derrida, Deleuze, Foucault, Lefebvre), as well as landmark achievements in science (Pasteur, the Curies, Lévi-Strauss). France's heritage of artistic, intellectual and philosophical resources contributes profusely to contemporary Western thought and if Americans dismiss the French as irrelevant it is most likely because they tend to view the world in black and white, up or down, winners and losers, whereas a more sensitive view includes many shades of grey. The French are famous for attending to the greys, and if their worldview deserves our attention it is not only because of their claims to distinction mentioned above, but in equal measure because the future course of world politics (and American politics as well) is better captured by half-tones and shadows than by the simplistic dichotomies that dominate the geopolitical imaginary in the US, the Middle East, and elsewhere.

Citizenship and Globalization

Citizenship has undergone many changes in conjunction with globalization. The networks of interdependency and identity that people depend on in day-to-day life are increasingly woven across political boundaries. As economic agents and cultural consumers people no longer occupy a geography of bounded places but rather a space of intermeshing networks and flows (Castells 1996). At the geopolitical level, we are increasingly cognizant of the dependency of the state on interstate alliances and agreements. Democratic institutions designed for the bounded state have been overcome and compromised to greater or lesser degree by these geographic scale transformations.

Some theorists are becoming aware of ways in which globalization has induced in democracies a democratic deficit (Agnew 2005, 221–3). First, the foundations of state sovereignty are undermined by global capital flows which force the opening up of markets and the rolling back of state protections of citizen health, security and welfare. Second, the territories within which authority is distributed under democracy are increasingly challenged by non-territorial manifestations of power such as international organizations and transnational corporations – world powers in their own right that often wield annual budgets greater than the GNP or GNI of the states where they are active. Third, the emotional purchase of the nation-state is forced to compete more and more with other bases of personal attachment. The allegiances forming personal identity are transferred from the nation to the region or to border-crossing identities like those enframed in the women's movement and the gay rights movement (Agnew 2005, 223), as well as virtual social settings –

communication environments – making up "flowmations," "ideoscapes," or what I call "bridgespace" (Adams & Ghose 2003).

These various shifts and transformations create a democratic deficit because traditional democratic representation is poorly suited to address, involve or represent citizens defined at multiple and overlapping scales and buffeted by forces largely beyond the control of the state. This detachment of social processes from the traditional foundations of citizenship causes citizens to become disillusioned with government insofar as they "are unable or feel unable to hold to account an institution that has law-making and tax-raising powers over them" (Meadowcroft 2002, 182). The US is caught up in this current as much as other countries, and although it has acted as a major instigator of economic globalization the US's continued affluence and even its security are threatened in significant ways by the forces of the global free-market (Agnew 2005, 189–218). Europeans, on their part, feel a loss of sovereignty from the distant super- (or hyper-) power across the Atlantic and from the EU, as the latter increasingly dictates state-level legislation and policy in Europe (albeit in accordance with the decisions of elected representatives and heads of government).

Nationalism probably accounts for much of the disparagement of the EU by its citizens. Nationalists can be defined as those who choose to structure their actions and perception around the territory of the nation, bracketing out awareness of chains of cause and effect that cross over national borders. Nationalists may persist in their attachment to the national scale even if their country is part of an international union, like the EU, or if their country dominates other countries militarily, economically, politically, or culturally, like the US. Neoconservativism of the variety that has recently come to dominate political life in the US promotes the narrowing of perception and commitment from the world or region to the nation-state, equating citizenship with the interests of business and consumption in the "homeland," and narrowing the space in which people are willing to imagine that they bear some responsibility for the problems of others. For example, wars for access to foreign resources can be fought in the name of democracy if one adopts a nationalistic American worldview without interrogating the nature of the American-backed system that generates conflictual forces.

Alternatively, citizens may choose to inform themselves of the chains of cause and effect that link actions across borders. This framing of identity suggests the cosmopolitan idea of moral obligations *of* and *to* distant Others. Cosmopolitans take the entire planet (including non-human life, in some cases) to be the society in which they act and the context in which they must assume (at least a certain degree of) moral responsibility. It is not impossible to be at once cosmopolitan and committed to one's national community. The concept of "rooted cosmopolitanism" (Appiah 1998; Cohen 1992) indicates that a country like France, with its liberal communitarian version of citizenship and overtly self-serving politics may still be cosmopolitan, albeit rooted in a national culture. The communitarian ideal is enshrined not only by the term "*fraternité*" in the French national moto, but also at times by "*solidarité*" as promoted by Léon Bourgeois, drawing on the philosophical foundation of Rousseau's Social Contract (Bernstein 1990, 225). France's expansive *mission civilisatrice* (civilizing mission) has coexisted with notions of mutual support and obligation to distant Others. Conversely, the US with its outward turned ideal of economic and cultural "free-flow" and its inward turned ideal of the "melting pot" may not be cosmopolitan if Americans understand international flows ethnocentrically and the purpose of "melting" is to assimilate diverse cultures to the dominant American norms.

Yet Americans are being forced by circumstances to confront the tension between bounded morality and the global system that supports their prosperity even as the French are being forced by circumstances to address their own tensions. On the one hand, France's laissez-faire attitude towards foreign political models (including those considered by the US to be so despotic or conflict-ridden as to call for military intervention) indicates a kind of nationalistic insularity; on the other hand, France's leadership role in Europe suggests the perceived inadequacy of this insularity over the long term as a basis for preserving French culture and promoting French interests. In short, many French citizens recognize that only by sacrificing a certain measure of their sovereignty can their nation resist incursions on its autonomy by global forces and American models of globalization.

Kant and Hobbes

A major ingredient in the neo-conservative worldviews prevailing in the post 9–11 United States is a "realist" or Westphalian geopolitical philosophy. This perspective assumes that the world is essentially a chaotic place and that force rather than diplomacy is the best guide to international policy. Two defining figures implicitly or explicitly drawn into arguments for and against realist geopolitics are Immanuel Kant and Thomas Hobbes. Kant's position on international relations is set forth in "Perpetual Peace: A Philosophical Sketch" (Kant 1983 [1795]). Kant views war between states as a natural condition but one that can be overcome by the establishment of treaties between democratic republics. He envisions a federation of sovereign states whose members treat each other with hospitality and who, for reasons of enlightened self-interest, choose not to engage in wars with each other. Hobbes (1985 [1651]) sought the means to overcome unrestrained and selfish competition through the absolute power of government, yet his antipodean vision of the "war of all against all" is what is most often indicated by the term *Hobbesian*. The distinction between Kant and Hobbes therefore has become in the current geopolitical debate a disagreement over the possibility of peace between nations and the feasibility of pursuing international treaties and agreements. For Robert Kagan (2002), an American neoconservative, Europe's "Kantian" perspective on geopolitics is a fantasy or delusion hatched in the protective embrace of the US. European commitment to international law and cooperation is the misguided product of Europe's failure to appreciate the significance of American protection because, he argues, it is primarily the United States which has made compromise possible and fruitful within Europe. Outside of Europe the world remains Hobbesian, in his view, and the US is sufficiently wise to recognize this fact and sufficiently benevolent to shoulder the burden of protecting its allies from each other and from the rest of the world. This account of the world system grants the US an exceptional status as the only country motivated by altruism rather than self-interest.

The Westphalian geopolitical model has been critiqued for its failure to take into account the plurality of voices within the state, the proliferation of non-state actors (IGOs, NGOs, multinational companies, etc.) and the motivations behind international aid (Dodds 2000, 37–51). Agnew describes the "territorial trap" of Westphalian worldviews as consisting of three erroneous assumptions: (1) that sovereignty requires bounded territorial spaces, (2) that societies correspond precisely to these state territories, and (3) that each state's gains inevitably come at the expense of other states (1998, 51). When promoted by the most powerful state in the global system, this view actually promotes non-cooperation among states since the axiomatic condition of realism, perpetual competition, implies that the dominant competitor can have no possible use for cooperation between states; interstate cooperation might lead to the formation of alliances of weaker states that would then be capable of truly competing with the dominant country. Mutual hostility and suspicion preserve the supremacy of the dominant state. What realist geopolitics dictates is a kind of global disorder that can be profitably maintained through the use of unilateral force and disregard for international agreements and organizations (Herod et al. 1998; Joxe 2004a, 2004b). So realist or Hobbesian geopolitics is the source of the democratic deficit perceived outside of the US, and that deficit is in turn a product of the undermining of global stability by American policies as much as the result of terrorists or "rogue states."

To restore meaning and relevance to citizenship, and to restore global stability, the bounds of citizenship must be pushed beyond the borders of the nation-state toward the construction of a transnational democratic system and its support structures. This does not imply that citizens would need to vote for leaders of other countries or for global rulers or administrators (although this would be a tremendous achievement). The democratic process consists of more than voting so its international element can be, and is already to a large extent, promoted through other means. Insofar as votes are the culmination of multimodal, multiscalar discussions that take form over months and years, and insofar as every vote brings a realignment of political forces which provides the basis for political discussion, then *being a citizen means participating in a process of formal and informal political discourse*. This discourse already crosses formal political boundaries and could potentially do so to a much greater degree. Political discourses are unbounded networks and their further expansion and elaboration via old and new media can only strengthen democracy in face of the democratic deficits incurred by globalization.

The Shaping of Opinion

The task of imagining the nation-state and justifying its global role falls, in particular, to those whom geographers Ó Tuathail and Agnew (1998) call "intellectuals of statecraft." These pundits and policy-makers are overwhelmingly dominant within the public discussion and debate of foreign affairs, defense and international relations. Such figures include presidents and prime ministers, their administrations, past and present government leaders and diplomats, key legislators, politically outspoken newspaper editors and columnists, and a wide range of media personalities from all segments of the political spectrum. Intellectuals of statecraft are the opinion leaders who set the terms of geopolitical debate: including and excluding issues from the realm of "legitimated" debate and building up associations that naturalize or "script" the nation-state.

Opinion leaders are complemented and often supported by the popular media acting as gatekeepers, determining what issues fall within the scope of legitimate debate and what issues are de facto beyond discussion. Todd Gitlin famously captured the political power of media in *The Whole World is Watching* (1980) in connection with news coverage of the 1960s social movements. He used the term "framing" to describe the way "symbol-handlers routinely organize discourse, whether verbal or visual" by applying "persistent patterns of cognition, interpretation, and presentation, of selection, emphasis and exclusion" (Gitlin 1980, 7). Framing is political insofar as biases and assumptions, or what Gitlin calls "little tacit theories," are always implicit in media representations, and such biases and assumptions control the public perception of "what exists, what happens, and what matters" (Gitlin 1980: 6). This ideological framing process involves ongoing inclusion/exclusion of people in the "we," the result of which is to encourage audiences to view Others as friends or foes, to support political and military endeavors, to accept certain interpretations of morality, civilization, decency and, at the most abstract level, to understand space and time in certain politically-charged ways (Johnson 2001; Dodds and Atkinson 2000; Sharp 2000: 126–7; Shapiro 1999; Laclau 1977: 100). Thus, the national image is often a product of the interplay between opinion leaders and the media, with multiple sites of representation and interpretation.

It is as much through representations of *other* nations as through representation of "our nation" that nationalist discourses construct the meaning of the nation. In the words of Gertjan Dijkink: "national identity is continuously rewritten on the basis of external events," and these events are understood in ways that are not strictly determined by events themselves: "foreign politics does not mechanically respond to real threats but to constructed dangers" (1996, 5). This indicates that nationalist discourse depends on a dialectical relationship between the nation and other nations as media and intellectuals of statecraft decide to represent the other nations, constructing them as threatening or harmless depending on their own needs and interests.

"Friendly" nations and states are, of course, constructed as friendly because it serves the interests of those in power and helps those elites maintain legitimacy.

When it no longer serves any purpose to represent a country as friendly, the "friendly other," the foreign ally, dwindles in public discourse. Gatekeepers in positions of power within media organizations decide whether it serves their interests to preserve motifs or change them and portray an allied country as an enemy, or vice versa. This is what the George W. Bush government decided to do with regard to its representation of France, for reasons that probably have more to do with currying the favor of a certain segment of the US population than with actual diplomatic relations between France and the US. Of course at the personal level, national imagery is not passively absorbed from public communications; people actively select certain meanings and reject others (Fiske 1987; Katz and Liebes 1985). Yet much of the time public opinion can be brought in line by the ideologies supporting nationalist scripts.

The current chill in US-France relations depends, as well, on the fact that the credibility of country X in the eyes of another country Y's citizens depends in large part on whether those in power in country Y find it *useful* to portray country X as friendly. The US's situation is particularly sensitive to foreign opinion owing to its unparalleled importance on the world scene and its symbolic centrality. A defiant stance in US government rhetoric reverberates due to this factor of utility so that foreign leaders quickly become antagonistic because it is useful to them to demonize the US (just as it is useful to those American leaders with nationalist backing to demonize certain foreign countries). US military and economic strength makes this polarization all the more likely since both of these forms of strength provoke resentment which can be turned into a potent political resource by intellectuals and leaders lacking in other assets.

The reverberation that characterizes the current period of international relations consists in the impossibility of limiting the audience of a comment to American voters and the utility of representations of the US as a disliked or hated Other. So France-bashing comments at a meeting of the National Rifle Association may well find their way into internationally accessible websites or onto radio talk shows in the US and abroad, then to newspapers around the world, and ultimately catch the attention of foreign dignitaries who are not always willing to look with indulgence on the bravado of American nationalists. The spillover between mediated communication contexts and the sharing of discourses is therefore central to the redefined geopolitical situation in which the US finds itself.

A shift of opinions in France from support of the US toward an attitude of neutrality or hostility is therefore of major theoretical and practical interest, not only for its own sake but as an example of a general process. A key question for the US is how quickly relations with such an ally can deteriorate. Fortunately, public discourses in France have yet to turn toward blanket condemnations of the US, or what could legitimately be called "anti-Americanism," despite symbolic targeting of France by American nationalists. And contrary to allegations, the French critiques of the US have been measured and targeted up to this point, focused on the President and his policies but also on the geographical roots of American neo-conservatism. This is part of the story I will tell. Another part concerns the media that illuminate different facets of a foreign event like an election.

Media Affordances

An English Professor at the University of London, Gunther Kress (2003, 3), demonstrates with unusual insight the way in which different communication technologies can affect the character of what is communicated. His analysis is useful because of the widespread misconception that technology is neutral and therefore outside the realm of political debate. The misconception derives not only from theoretical frameworks that free human agency from external determinants, but also from philosophies such as Marxism that focus on one form of determination (economic, for example) at the expense of all other forms of determination.

Kress notes that certain messages are more easily conveyed depending on the medium of communication. If using words as one's communication medium, it is necessary when representing something with parts to relate each part to the whole in terms of possession. For our purposes, the best example is the linguistic framing by which a nation *has* X number of people, the "has" implying possession. If we communicate about things by drawing or mapping them, in contrast, we do not link the part to the whole in terms of possession, but rather in terms of containment, pattern and position. Mapped nations have shapes and their population is clustered here and there, in cities and towns. In short, pictorial representation forces the communicator to make commitments regarding *position* while verbal communication forces the communicator to make commitments regarding *possession.* Different kinds of commitments in turn afford different meaning-making opportunities to communicators, a quality of each medium which Kress captures in the term *affordance.* We will be considering scholarly discourses, television, newspapers, and the Internet in this book – four media for which the affordances are quite varied.

The role of newspapers in perpetuating international political debate is not the same as that of television or the Internet, or of physical political gatherings like caucuses and political conventions. The proliferation of media therefore implies a proliferation of the spaces of political debate – a fragmentation of the "public sphere" (Habermas 1991). If the public sphere is fragmenting, and if people are able to engage with distant Others through an increasing number and range of media, then it follows that as citizens people are no longer attached to a single space of citizenship, or even to a set of nested spaces of citizenship (Herb and Kaplan 1999), but instead they occupy the intersection of many physical and virtual political spaces. These spaces are experienced and valued in ways that are qualitatively different. For example, one is less committed, generally speaking, to the places one accesses via television than to the places one accesses via telephone. From this it follows that media are not politically neutral but, as Gunther Kress argues (2003, 5), they "change, through their affordances, the potentials for representational and communicational action by their users," an idea in keeping with the observation that "technologies are contingent social products that influence and help constitute systems of social relations" (Sclove 1995, 61). So every place is subject to varying political dynamics of the media impinging on that place; it lies at the intersection of many circles of loyalty and political affiliation.

This image is not original but was first supplied by David Ben Gurion and appropriated by Mitchell Cohen (1992) to demonstrate his useful concept of "rooted cosmopolitanism." In Cohen's words (1992, 482), "to stand in many circles is to accept the principle of plural loyalties," and while his conception of these circles, like that of Ben Gurion, is defined in terms of social collectivities defined by nationality, class, religion, and language and so on, the proliferation of communication technologies makes available a vast number of different socio-spatial configurations defined by media affordances and particular politics. As Poster argues (2001, 4), media are not simply tools used by "the existing cultural figures of the self – race, class, and gender, or citizen, manager, and worker" but they in fact serve as distinct means for the reconstitution of social configurations and corresponding participants, therefore each medium functions "as a space that encourages practices that, in turn, serve to construct new types of subjects." To develop the model, then, subjects are not only positioned at the intersection of various communities and polities; they also experience the existence of roots to various (and often distant) places that crystallize particular political ideals and agendas. Communication lifts people out of their immediate social environments and re-grounds them in selected social environments – the flowmations, ideoscapes and sign/symbol systems that form each person's media environment. In other words, media deterritorialize and reterritorialize in a great variety of ways (O'Tuathail and Luke 1994). Thus while one may use newspaper, radio, or e-mail to construct a Republican, Democratic, Social Democratic, or Green worldview, the various media in fact highlight and obscure different elements of these attitudes and worldviews, relating audience, nation, and world in distinct ways.

Contrasts in Citizenship

Not only do communication technologies rework territories, but the act of communication itself is constitutive of here and there, us and them. When Louis XIV declared "*l'État c'est moi*" (I *am* the state) he was saying something about himself and about the French state, of course, but also something about his listeners. He was addressing his audience, the French people, as subjects, in effect reminding them that they were nothing more than subjects (not citizens). Three-hundred and fifty years later Christian Saint-Étienne (2003) raised the battle cry "*La puissance ou la mort*" (power or death) when considering the future of Europe vis-à-vis the United States. To make sense of his appeal for European unity requires, first, that we understand it as an appeal *to Europeans,* that is to say an audience that can imagine itself as primarily European in character or essence (rather than primarily French or Belgian, for example). Second, we must understand it as an appeal to *citizens* as opposed to subjects, that is, *active participants in the legitimation of the political power under which they live.*

To see all this we must understand that "subject" and "citizen" are distinct and markedly different ways of understanding the individual's role in the world, and the individual's essential tasks with regard to his or her fate. The English word "citizen" derives from Old French *citeain* (which became *citoyen* in Modern French) meaning a freeman of a city or town. *Citeain*, in turn, derives from *civis* or *civitas*, a member of a Greek *polis* who was entitled to participate in the polity, to affect the conditions under which he (at the time women were not citizens) was ruled. This in turn carried the notion of rights. *Bourgeois*, for its part, originally referred similarly to someone with certain legal rights under a feudal lord. In Marx and subsequent authors "bourgeois" came to mean the conventional, middle class and materialistic values of capitalists while "citizen" came to mean a member of any group, organization or collective – in effect, a person with rights. Two competing ideals emerged: one the bourgeoisie whose greed generated affluence, the other the citizen whose voice ensured justice, and the bourgeois ideology created obstacles to universal citizenship.

The consciousness of American citizenship has evolved during the past century as a consequence of American geopolitical dominance – the dominance that Henry Luce captured with the term "American century" in 1941. While Luce's vision included a sense of global responsibility "to be the Good Samaritan of the entire world" it is the other part of his program, "to determine whether a system of free economic enterprise…shall or shall not prevail in this century" that has won out by the 21st century. Americans have promoted *bourgeois* ideologies above and against the ideals of democratic citizenship. John F. Kennedy followed his inaugural appeal to "My fellow Americans" with an appeal to "My fellow citizens of the world," but no subsequent American president has felt inclined to do the same. The announcements, pronouncements, and arguments of American leaders, military strategists and political theorists seldom articulate transnational notions of citizenship today. The interdependence of corporate capital and political power in the US has produced a master narrative shared by economic and political elites alike, a geopolitical script in which the global free market system that enriches the US through the labor and resources of poor countries is taken as proof that the US must be a Good Samaritan. Coupled with this miraculous reconciliation of greed and generosity is the equation of interventionist foreign politics with "good" American citizenship. In Luce's words:

> America cannot be responsible for the good behavior of the entire world. But America is responsible, to herself as well as to history, for the world environment in which she lives. Nothing can so vitally affect America's environment as America's own influence upon it, and therefore if America's environment is unfavorable to the growth of American life, then America has nobody to blame so deeply as she must blame herself. In its failure to grasp this relationship between America and America's environment lies the moral and practical bankruptcy of any and all forms of isolationism.
>
> (Luce 1941)

This vision deliberately obscures the connection between sovereignty and citizenship. It transforms the multiple spaces of citizenship worldwide into an "environment" of US citizenship. In so doing it *overbounds* the sphere of American citizenship. In part this overbounding is achieved by strategically invoking religious overtones.

From "manifest destiny" of the 19th century to Luce's 20th century invocation of the Good Samaritan, to Cold War constructions of the Soviet Union as the "evil empire" (Sharp 2000), to George Bush's concept of an "axis of evil," a metaphysical, sacred order has been invoked repeatedly to justify American hegemony and control of its global "environment." This rhetorical strategy in effect denies citizens of other countries the right to self-governance, leaving a more limited concept of "democratic" participation instantiated as the obligation to participate in a game the rules of which are determined primarily by the US.

For the French, citizenship takes a different form. Historically, it emerged in opposition to an *Ancien Régime* that constituted people as subjects in a feudal order defended and legitimated by the Church so it is not surprising that churches were ransacked during the French Revolution. Even today monarchists are described by one observer as "Christian fundamentalists" (Bernstein 1990, 201). To be a French citizen today is accordingly to belong to a *secular state,* not merely by historical contingency, but by grace of the democratic struggle linking citizenship to the dismantling of a Church-backed feudal regime. Charles Conte demonstrates that a secular notion of the polity is emblematic not only of the French view but also of a more general, European view of democracy:

> Europe is post-Christian. Certainly the diverse Christian faiths still count millions of adherents, but Christianity is no longer the intellectual and moral framework of thought and action for Europeans. It is in France that the general trend has culminated most clearly in the legal secularization of state and schools. To ensure the freedom of conscience for every citizen (believer or nonbeliever), the Republic refuses to recognize, sponsor, or subsidize any religious organization. Its secularizing legislation remains, despite its imperfections and challenges, a model. Guaranteeing to each person true freedom of choice has made a more authentic citizen than in states where a religion or religions are privileged.
>
> (Conte 2004, 126)

French citizenship is an active mode of national belonging, grounded in a space the state creates for individual choice, which in turn depends on the exclusion of religious symbols from public forums in general and from state-sponsored institutions in particular. It is notable that France, "often pointed to as a zone of changelessness by those skeptical of claims about an emerging world of flows challenging that of territories" (Agnew 1998, 57) has taken the lead in developing a European monetary system and planning for a unified European defense structure and political system. But if the scale of France's geopolitical imagination is changing, its fundamental political attitudes are not. Along with secularism France demonstrates a fondness for political debate, reflected in a broader political spectrum (more parties, more extreme parties) than in the US and in the higher incidence of labor struggle than in the US.

From a French perspective, certain elements of the US political community are archaic and flawed as they fail to separate the silencing power of religion from the communicational spaces of civil society and as they favor bourgeois values over citizenship. The French notion of citizenship draws more on the communitarian tradition in which one is a citizen by virtue of participatory membership in a community rather than simply by virtue of being born in a particular location. French citizens are likewise expected to sacrifice more than American citizens for the sake of communal goods such as equality, fraternity, and solidarity. While these terms are abstract, at least to Americans, the implications can be very concrete. The obligation to preserve public educational spaces as separate from non-secular sources of power leads, for example, to the prohibition of Muslim headscarves in schools (Thomas 2006).

The French model is far from pure communitarianism, such as that of a monastic community or tribe, but it is a type of "liberal-communitarianism" (Delanty 2002) in which citizens overwhelmingly support institutions that protect the general welfare, such as universal medical insurance, and ascribe to the institutionalization of certain standards of comfort such as paid vacations and a thirty-nine-hour workweek. Along with these protections and comforts comes greater willingness to accept taxation for such purposes: "In France today, by contrast [with the US], Socialists and members of the RPR, the main conservative party, believe essentially in Big Government" (Bernstein 1990, 251, 225–6). This economic view has its political correlate: esteem for the state that seems in the final analysis to cross into a kind of veneration (Bernstein 1990, 279–10), as if the French state occupied the niche left by the Catholic Church it supplanted. French liberal communitarianism seeks to "anchor political community in a prior cultural community," taking that entity as a given and conceiving of the self as "always culturally specific"; the erosion of national culture would imply a loss to oneself (Delanty 2002, 163). Of course French political views vary from Left to Right (we must not forget that these terms were first given political connotations in the city of Paris!) and the views of community vary accordingly.

Nonetheless France's political middle-ground lies farther to the Left than in the US. Citizens in this liberal-communitarian polity often perceive America's free-market liberalism and political conservatism as threats to the foundations of citizenship. More specifically, the French are threatened because the US employs a Hobbesian or realist geopolitical framework and because the American social model upholds fragmentation, commercialization, and self-interest, rather than maintaining the ideal of a community. This value structure threatens the French notion of citizenship, embedded as it is in a state that protects and nurtures the individual while imposing on him/her certain abstract and concrete duties to the community.

Conclusion

Fall 2004 was a period of international/interstate communication. French opinions about the American presidential election of 2004 serve as a snapshot of a small component of world opinion. Owing to the globalization of the public sphere, these opinions are not simply idle chat from the sidelines, but elements of the reconstitution of citizenship at the global scale. Such constructions include many elements that are not specific to one particular country, such as the perception of a democratic deficit caused by the forces of international trade and the overwhelming power of the United States. A sense of the loss of national sovereignty is not specifically French, nor is it separable from American politics either as a cause or an effect. In other words, the snapshot of French geopolitical discourses reveals patterns that are seen in many other places and contexts at this period in world history.

The French observers do serve as an interesting case, however, because of France's role in global politics and the longstanding relationship between the US and France, which, as we will see in the following chapters, is a troubled and uneasy alliance. Furthermore, the French people have multiple affiliations, many of which are border-crossing identities. They are constructed as transnational citizens and their discourses cannot be understood within the territorial trap of Westphalian, realist geopolitics. Rather, they provide insight into how American politics is being perceived by ordinary people and intellectuals of statecraft in secular, democratic, left-leaning societies. As such, they provide a lens on European views, in general. They also show how the various media affordances highlight different elements of the geopolitical situation.

We hear from influential conservatives such as Samuel Huntington (1997) that the best response the US can take to external critique is internal discipline – a policing of boundaries between Us and Them: "The futures of the United States and of the West depend upon Americans reaffirming their commitment to Western civilization. Domestically this means rejecting the divisive siren calls of multiculturalism" (Huntington 1997, 307). Whereas the Monroe doctrine restricted American interests and obligations to the Americas,[1] Huntington would now restrict American interests and obligations to the "Western" world including Europeans but excluding Latin Americans, whether south of the US-Mexico border or living in the US. Clearly Huntington's "West" is defined more by affluence, race and ethnicity than by "culture" or "civilization," even though he appears to base his argument on the latter two terms. Huntington's pessimism regarding the potential of non-Western "civilizations" (read poor, nonwhite people) is balanced by his optimism regarding the future relations between the US and Europe, which will supposedly "become more and more aware of their common Western cultural core that binds them together" (Huntington 1997, 307). He is implying that Europeans will soon share America's xenophobia, and that they will reverberate in a positive and harmonious way with bourgeois ideals, relinquish a communitarian notion of citizenship and

1 President James Monroe (1823) staked out North and South America as a United States domain off-limits to the rest of the world and particularly, "not to be considered as subjects for future colonization by any European powers" while US policy in regard to Europe was explicitly: "not to interfere in the internal concerns of any of its powers; to consider [each] government de facto as the legitimate government for us; to cultivate friendly relations with it, and to preserve those relations by a frank, firm, and manly policy, meeting in all instances the just claims of every power, submitting to injuries from none" (Monroe 1823).

the state, and cling to the worldview of Westphalian realism – all of this despite globalization and their very different role in globalization from that of the US. In the following chapters Huntington's miscalculations will become apparent and that the source of his error is not, for the most part, an upsurge of "anti-Americanism" but rather a peculiar inversion of power in which the most powerful country in terms of money and weapons is unusually vulnerable to the power inherent in words, images and ideas.

Chapter 2

Geopolitical Representation and its Contexts

> It is because I am unable to see at once all that is around me that I am allowed thus to select and separate the objects of my choice from among so many others which it pleases me to contemplate.
>
> de Tocqueville, *Democracy in America*, p.704

In democracies, the government policies of tomorrow arise out of the grassroots communications, mass media representations, and intellectual debates of today. When we speak of the US, France, or any other nation we must keep in mind that the nation is not an object, something that simply *is*, but rather a process of on-going representation. It would consequently be foolish to calculate the future of any country purely in terms of its military and economic strength while overlooking the ideological and discursive factors likely to shape its future. Representations of nations constructed and disseminated in and by other nations – newspaper articles about trade partners, written debates about international relations, television documentaries about foreign wars, and so on – are fundamental aspects of the symbolic process of national identity construction. In a world where democracies are becoming more common, global public opinion increasingly shapes the fate of nations; and every country's international reputation matters. The most powerful country in the world is subjected most thoroughly to this rule. The geopolitical situation of the United States – its ability to maintain security, stability, and dialogue with other nations – depends much less over the long term on how Americans see themselves than on how the rest of the world represents and conceptualizes "America."

I use the term "America" in this book to indicate a representation of the US. In keeping with this nomenclature, *l'Amérique* is used at times for "America" as it is represented in French discourses. The implication of this nomenclature is that there are many "Americas" and these various Americas – positive and negative, powerful and weak, rising and declining, authoritarian and democratic – are not just representations but forces of a political nature circulating in political space. Therefore, the entity that engages in wars is the *US* while it is *America* that is represented in domestic and foreign discourses *about* those wars. These Americas are not "merely" imaginary, although they are imagined. They become real through democratic and diplomatic processes and their reality is a fundamental constraint on US geopolitical relations. In keeping with a process noted by Katz and Dayan in another context (1985), fantasy images sold to Americans as a means of escaping from their everyday reality are often understood by foreign viewers as windows onto "the way it is" in America. America disseminates these lurid fantasies at its own risk.

In short, the nation and the concept of nationality are ongoing imaginative constructions and major elements of nationalism are constructions of the Other (see also Anderson 1983; Appadurai 1996; Dodds 1996). These representations naturally evolve through time as do the representations of one's own nation. Representations, images or visions of one country by another country say as much about the country that is doing the depicting as about the country that is depicted. The improvement in American portrayals of Japan (and vice versa) between 1945 and 2000 illustrates this point. Representations of nations by the people of other nations can change from positive to negative or vice versa depending on economic, political, and cultural conditions in both countries. The most notable shifts in international outlook occur before and after key events in the history of societies like the fall of the Soviet Union or the attacks of September 11. Sometimes representations mirror perceptions; what is said or written in a country reveals what is actually believed or perceived. At other times, more perplexingly, national representations play up certain conflicts and play down other conflicts in ways that mask the real levels of international antagonism. An illustration of this phenomenon is the fact that France is one of the US's most vociferous critics yet the US-France alliance is one of the longest standing international alliances in the world.

This chapter examines the idea of geopolitical representation, beginning with the particular geopolitical representation in question here – America as it is viewed and understood from without – then progressing to the dynamic interrelations between geopolitical representations, conceptions and perceptions. The construction of geopolitical motifs is linked more clearly to geographical context than has previously been the case. We then turn to Joseph Nye's useful distinction between "hard power" and "soft power" as a way to interpret the role played by geopolitical motifs in the "real world" and as a caution regarding the neglect of geopolitical representations.

Constructing America: Many Visions One Hyperpower

In the words of one French observer, "identification of the United States of America as an empire is, at the dawn of the 21st century, the single most shared thing in the world" (Roger 2004, 107). This sentiment is one thing that unites France and the rest of Europe: "A collective apprehension about the United States seems the only glue that binds Europeans together. Scathing stories about the United states' death penalty, shootings in high schools, unforgiving market, and lack of welfare abound in the European press" (Riotta 2000, 87; quoted in Nye 2002, 33). The former French Foreign Minister, Hubert Védrine, has coined the term *l'hyperpuissance* (the hyperpower) to indicate the US's dominance of the world in economic, political, cultural, and military spheres, and this dominance goes a long way toward explaining not only the global preoccupation with the US but also the anxiety, fear, and loathing that the US has incurred. For better or worse the US is increasingly not just an other nation but *the* Other nation, a point of reference that helps non-Americans make sense of who they are as citizens and individuals. Responses to the US in the form of symbolic representations and corresponding foreign policies will, in turn, affect the opportunities and constraints available to US policy makers and ordinary Americans traveling or doing business abroad.

Thus, whether we study political discourses in China, Saudi Arabia, France, or Quebec, a central place within the respective fields of discourse is occupied by "America," providing keys to the worldviews in each country, the prevailing perceptions of past and future, and the visions of that state and/or nation within the world order. For better or worse, America now stands as an extreme or an absolute, an icon and an ideal, sometimes positive but increasingly negative, that helps people make sense of the limitations, constraints and imperfections of the places they inhabit, as well as the changes their surroundings are undergoing, and the nature of the world as a whole. For some observers, America stands for a kind of utopia. The utopian vision diffuses globally as the collective dreams of American society – rendered on film and video, in popular music and on the Internet – link certain iconic landscapes like the desert Southwest or New York skyline to myths of individualism, adventure, and the opportunity to succeed. For other observers, America rots from within, spreading contagion in the form of moral depravity, crime, violence, injustice, and secularism – visions based partly on different readings of the same American media products that generated the positive images mentioned above – giving signs of America's imminent collapse if only it can be resisted strenuously and courageously. Of course it is largely through US media products that the world comes to know America and different Americas appear to the world due to the imposition of different interpretive frameworks by various observers on the vast outpouring of heterogeneous American cultural goods (Ang 1985). The result is a range of habitual attitudes linked to nationalistic discourse and indirectly to perception. One viewer watches a presidential inauguration and sees "America" while another watches the same footage and sees "Republicans"; the difference lies in the viewer's subject position, with those outside the US readily interpreting any particular sign or symbol from America (in any communication medium) as a *synecdoche,* that is, as a part of America that stands for and captures the essence of the whole.

The US, viewed as a sovereign state, occupies a clearly defined territory on the earth's surface, but America, as a symbol – of a society, a culture, a way of life, or an ideal – constantly transcends those geographical borders, penetrating to the most remote regions of the planet. Complementing America as an imagined place is a temporal "location" for the American culture that is sometimes the future but often, in a way that would surprise most Americans, the past. For some foreign observers, America appears as an ideal tomorrow of luxury, even hedonism, but for many others the US is trapped in a yesterday of outmoded lifestyles and attitudes – violent, superstitious, and antisocial – a vestige of 19th century rural frontier that is distinctly out of place in 21st century urban world. American attitudes about guns, abortion, homosexuality, the death penalty, and the role of government are all reasons for this less flattering interpretation which is particularly prevalent in France. Meanwhile, US Defense Secretary Donald Rumsfeld has branded France and Germany with the derogatory label "old Europe." The term is shorthand not only for their date of EU membership but also for their "failure" to view American dominance and unilateralism as the way of the future.

Of course the US and Europe exist at the same moment in history. In the present the two places are intensely interdependent. Their differences in power, culture, and experience inspire different social trajectories, so that each national culture (judging by dominant discourses) sees itself moving into the future while the other lags behind or moves backwards. There is a fascinating and disturbing kind of cultural relativity at work here.

The symbolic construction of America is steered by certain persons within and beyond US borders whose voices resonate with particular audiences, responding in particular ways to their audience's sense of place and history, their sense of community identity, and their personal identity. These are the intellectuals of statecraft and the media executives. In the terminology of media studies, they are the "opinion leaders" and "gatekeepers" of a society (Lazarsfeld, Berelson and Gaudet 1944; Westley and MacLean 1957). Their actions to filter and shape communications attempt to construct a worldview in which the "imagined community" that is the nation (Anderson 1991) is rendered in self-justifying, seemingly natural terms. For this it may be useful to attack the symbols of a different or competing power – and America fulfils that role *par excellence*: as an attentive and responsive democracy America lends support to more progressive elements of foreign societies but as a mercenary and arrogant world power America provides symbolic resources for antidemocratic leaders but also clashes with the true proponents of democratic ideals.

In this book I will draw on the insights of critical geopolitics (O'Tuathail and Agnew 1998; O'Tuathail, Dalby and Routledge 1998; O'Tuathail 1996; Sharp 2000; Shapiro 1999). My project deviates somewhat from the dominant emphases in critical geopolitics, however, because the texts I examine are ambivalent about the powers in question (the US, France, Europe). They are not intent on the scripting of a particular form of state domination (e.g. O'Tuathail 1992) nor on the rhetorical attack of state power (e.g. Dodds 1996), but on something between the two and more complex: the representation of a place deeply inscribed with power (the US) but complicated by geography (regional variations in political culture), history (the emergence of terrorism), and culture (widely divergent visions of America by Americans themselves). In the French media, the "America" of 2004 is, quite simply, a complicated place, internally divided into camps and regional factions, as well as into urban and rural contingents, with some voices that echo and are echoed by the French observers, and other voices wholly alien to the French. I find that simple anti-Americanism is generally avoided in these representations of the US from across the Atlantic although unreserved praise of American policies is extremely rare, in part because it is hard to direct praise towards a recipient which receives praise and criticism with equal aplomb. Consideration of these geopolitical discourses draws us deeper into America's dilemmas, and into France's perplexity about America, and into the representational dynamics of globalization.

Geopolitical Triad

Gertjan Dijkink (1996) uses the term "geopolitical vision" for the way one country perceives and represents another country; the study of such visions addresses "any idea concerning the relation between one's own and other places, involving feelings of (in)security or (dis)advantage (and/or) invoking ideas about a collective mission or foreign policy strategy" (1996, 11). As the global system of states responds to the newly aggressive US policy of the post 9–11 period, and this multiple re-balancing act interacts with other geopolitical factors ranging from ethnic separatism to the rise of "world cities," from terrorist networks to emerging forms of supranational cooperation like the EU and FTAA, attention to geopolitical visions is tremendously important.

Within the larger subject of geopolitical discourses an important area of study concerns "official and public reactions in different countries during and in the wake of a critical international event" (Dijkink 1998, 198). Events alone do not shape the political landscape; their power comes only as they are transformed into *motifs*, structures of meaning that make sense of the world. These structures are called "frames" by Todd Gitlin (1980), "mythologies" by Roland Barthes (1972, 116–17), and "codes" or "geopolitical visions" by Dijkink (1996; 1998). I prefer "geopolitical motif" to the term "geopolitical vision" because seeing, vision, and surveillance are so often associated with dominance and control (Hillis 1996; Virilio 1989; Olwig 2001; King 1997). It could be argued that "geopolitical idea" is more to the point, but a vision (a way of seeing something) is not the same as an idea (a way of thinking about something), which is again not the same as a motif (a way of representing something). These terms are different but interrelated and we gain something by reminding ourselves of the differences and exactly how they are related.

We could start by describing the relationship as follows: perceptions shape conceptions, which in turn shape representations. We see X happen (perception), our worldview changes to include the possibility of X happening (conception), and we speak to others about X happening in the past or present, and about the likelihood of X happening in the future (representations). Thus the formula below is one set of relations between these terms:

perception → conception → representation

By the same token, if we believe in advance that Y can happen, that belief may lead us to speak of Y happening, and our words, affirmed as valid by our listeners, make us inclined to see Y rather than Z when we look at the world's complicated and ambiguous events:

conception → representation → perception

What we need to reconcile the triad and its complex relationships is a model that incorporates all forms of interdependence between representations, conceptions and perceptions (fig. 2.1).

The triad exists in a particular geopolitical context, as well. The reality of geopolitical context – conflicts, treaties, relative power, and so on – rises above any of its elements and involves objective technological, material, spatial and historical conditions – the richness of history and geography. Some nations place a greater emphasis on boundaries than others; some are fragmented while others are unified; some are relatively isolated while others lie at busy geographical crossroads. The internal and external geography of the nation play important roles in geopolitical perceptions, conceptions and representations. The US is not represented as similar to the Gambia, for example, no matter who is doing the representation.

Dijkink refers to representations of external geography as the "geopolitical code" (1996, 12) and explains that this code consists of the way other nations/states are constructed as enemies or allies, models to emulate, competitors, museums of the past, prognostications of the future, disaster zones, or sources of raw materials to import, to list some common interpretations. But here again the terminology could perhaps be more clear, since a code is a rule for translation of one representational system into another, while "motif" serves better to capture the idea of a theme with many variations.

Geopolitical motifs like Barthes' mythologies, give every geopolitical image or text a meaning that "is *already* complete, [that] postulates a kind of knowledge, a past, a memory, a comparative order of facts, ideas, decisions" (Barthes 1972, 177). A motif can be evoked by a wide range of media and by various "languages" embodied

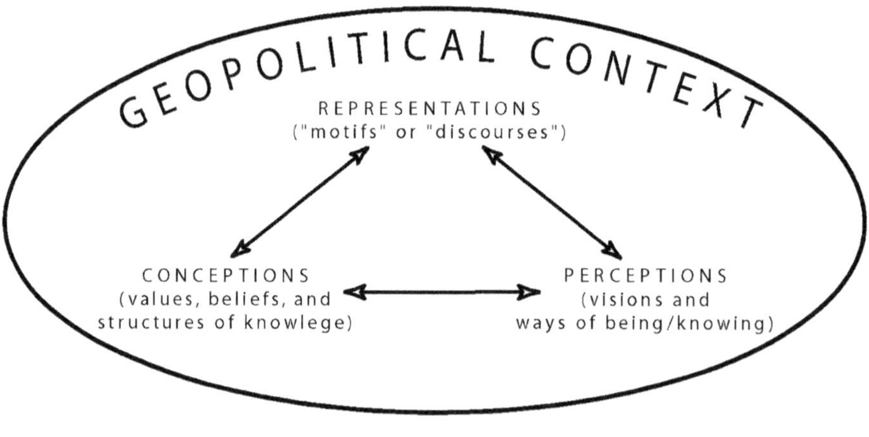

Figure 2.1 Geopolitical representations in their context. Representations relate dialectically to both conceptions and perceptions, and the triad of representations, conceptions and perceptions is shaped by its geopolitical context. Author's diagram.

in the media (photographs, government proclamations, news reports, internet posts), which a particular motif "gets hold of in order to build its own system" (Barthes 1972, 115). In essence a motif is parasitic on other symbolic systems, but how exactly does this parasitic function work? Todd Gitlin supplies the answer by defining media frames as "principles of selection, emphasis and presentation composed of little tacit theories about what exists, what happens, and what matters" (Gitlin 1980: 6). Gitlin's use of the term "little" reminds us of the sketchy and impressionistic nature of conceptions, of their simplicity and dependence on stereotypes even when they seem solid and self-assured. By calling them "tacit" he also reminds us of the fact that although conceptions may occasionally rise to the surface as in President Bush's motif of the "axis of evil," more often they circulate below the surface. Their simplicity is intentionally deceptive. Representations often work like the fact-filled *Reader's Digest* story that "both aestheticizes and sanitizes contentious issues and deflects attention away from deep or critical understanding" (Sharp 2000, 37).

These observations help us to understand how it can be that France and the US, while generally maintaining a relationship of uneasy alliance rather than enmity, nonetheless trade derisive remarks on a regular basis. French leaders find it useful to critique the US while American leaders (at least during the George W. Bush administration) have found it useful to cast aspersions on France. French observers critique the US much more than they critique Germany (a difference at the level of representation), yet a sense of mild distrust shapes French views of both the US and Germany (a similarity at the level of conception). Why the difference? Simply stated, distrust of Germany (a conception) is handled more effectively through rhetoric of harmonization (a representation) because such harmonization is possible and is already occurring within the framework of the EU (Keiger 2001, 216–17). American power cannot be tamed or neutralized (a perception) by a regional body (NATO having demonstrated this point to the French), so favorable representations of America appear to serve no purpose. Critiques of the United States, in contrast, maintain a sense of dignity and national pride in face of US power and rhetoric while at the same time supporting the Europeanist ambitions (a conception) prevalent among the French leadership (Brenner and Parmentier 2002, 15–18). Within France's geopolitical context, then, criticism of the US has *utility* but criticism of Germany does not.

US rhetoric must also be decoded in the context of American goals and objectives. Many in the US would not like to see Europe develop a strong and independent military force because the US currently enjoys a monopoly of military power. It nonetheless serves the interests of US politicians and intellectuals of statecraft to criticize European countries for being weak and dependent (e.g. Kagan 2003). This kind of critique (including the representational motifs, *we are doing a job for the world* and *they selfishly refuse to help*) serves as justification for American military aggression by sidestepping larger (conceptual) questions about whether the "job" needs to be done or is morally justified, as well as who is served and who suffers as a result of this "job," as well as why the US has the greatest means at its disposal for the job. Each country's geopolitical context shapes its rhetoric (representations), understanding (conceptions) and worldviews (perceptions) and the triad in turn shapes its evolving geopolitical situation.

According to Dijkink (1996, 2), "Living somewhere means being exposed to the continuous stream of discourse produced by a local society *and* experiencing events which differ in kind from those happening elsewhere in the world… Each place suggests models for the world and provides blind spots as well." While every place contains multiple and conflicting motifs, it is also true that the number of geopolitical motifs structuring discourses in a particular place is always finite. Therefore one's place, at any given time, has an important role to play in shaping the meaning of one's experience (perception and conception) of other places, and geopolitical motifs are best understood as *place-bound, simplistic theories about history and geography that usually remain implicit rather than explicit, but that attach themselves to a huge number of discourses.*

Our current understanding of geopolitical visions/representations/conceptions clearly owes a debt to Said's *Orientalism* (1994) which laid bare the colonial mentality behind both criticism of and fascination with "the East." But, as Dijkink points out, this seminal work is flawed. Dijkink asks how identifying constructions of the "oriental" permits one: "to delimit another world (Europe) and identify it with a systematic field of knowledge about the East?" (1996, 8). What he means is that for Said Europe appears to be an undifferentiated space from which geopolitical representations are diffused, and this undifferentiated space of the European dominating vision is itself a perception, conception and motif of "the West" no more real than "the Orient." In short, Said's critique (a representation) itself employs a stereotyping gaze (perception) that obscures the diverse array of geopolitical visions Europeans have demonstrated.

A more nuanced approach to geopolitical discourse delves into geopolitical representations of and by particular nations. The national scale is generally the most appropriate scale at which geopolitical visions can be identified and analyzed. As a group with a common language, heritage, and various other cultural characteristics, a nation is likely to embody a fairly homogeneous set of geopolitical visions. Only when a nation is essentially coextensive with a state does it make sense to use a state as the framework for analysis of geopolitical representations.[1] The boundaries of the French "hexagon," for example, demarcate both nation and state (although Corsica and the overseas *départements* are slight disruptions of this pattern). Therefore the geopolitical motifs of French media at a time like an American presidential election should be sufficiently coherent for the purposes of analysis. Nevertheless, we should not lose sight of the fact that these motifs are differentiated "internally" along the classic social fracture lines of ethnicity, income, education, gender, and locality.

1 The only significant flaw in Dijkink's intriguing *National Identity and Geopolitical Visions: maps of pride and pain* (1996) is the acceptance of the state borders as the framework defining particular geopolitical visions because such visions are often associated with nations strikingly different from the existing bounded states, as in the case of Québec or Iraq. It is not surprising therefore that Dijkink finds Iraq to be "totally lost."

Nationalism and the Triad

Public discourses lend symbolic power to various social entities ranging in scale from the nuclear family to the community of all living things. Each scale of social organization offers peculiar resonance to the self-image of the perceiver, the "audience member," and does so in ways that are deeply embedded in culture. The nation's particular position is that of an identity foundation between spatial extremes. Moving down the scale we encounter the local community – the town or neighborhood. Moving up the scale we encounter the regional society such as North America or Europe. Each scale is only as important as people make it out to be through discourses. A resident of Cherbourg, France, may identify primarily with her local region, Normandy, or with France as a whole, or with the international region of Europe. She may also identify with western (*occidental*) culture, or with the world as a whole. Most people in fact identify with several levels of social organization at once – developing what Herb and Kaplan (1999) call "nested identities." Each scale of identification offers a person something different so, for example, a sense of historical roots and cultural belonging may come from the region while the state offers political power and a common market. Then again, the market may be perceived as regional in scale, as with the EU and one may feel historically rooted in the nation of the Netherlands, Germany or France.

Opinion leaders who wish to "script" the nation for a range of differently situated audiences generally try to engage with various forms of sectionalism, subordinating local loyalties to patriotic sentiments and at the same time asserting the primacy of the nation over the broader transnational region as a framework for geographical affiliation (O'Tuathail 1992; Herod, O'Tuathail and Roberts 1998; O Tuathail, Dalby, & Routledge, 1998; Agnew, 1998). Nationalists attempt to redirect or "claw back" (Fiske and Hartley 1978) discourses that downplay the nation in comparison to larger or smaller scales of identification; that is, they symbolically suppress both local and transnational identities. Nationalistic discourses always address people as members of a nation, thereby working to reinforce a national scaling of the listener's identity, over and above other ways of situating oneself in the world. Nationalist discourses may be about many different things on the surface (development, national security, art, science, etc.) but below the surface they have at least one thing in common: the *interpellation* of each audience member as a citizen of a particular nation (Sharpe 2000; Laclau 1977). By accepting, affirming, applauding or joining in nationalist communication, one automatically accepts a particular construction of one's own identity as a "countryman." To listen is to be nationalized, although one may hear and reject what one hears. Sub-national and supra-national identities can be mobilized by the intellectuals of statecraft if doing so serves the interests of the nation or the party involved, but the mobilization of other scales of identity is framed by the overarching nationalist objectives, and often interpellates the listener as, above all, a person of a particular nation. In other words: nationalisms speak "spatial languages" other than nationalism but their master discourse is the nation as the foundation of self-identity.

We can trace the sources of any geopolitical motif to five interrelated aspects of culture that shape a nation's interpretive strategies (Dijkink 1996; 1998). First, geopolitical motifs reflect key moments in the nation's past: victories, defeats, civil wars, revolutions, the signing of important documents like treaties and constitutions, and so on. Every new event is interpreted through the lens provided by past experiences. Second, geopolitical motifs embody the concept of a national mission that positions the nation in regard to an imagined future. Most often the nation is seen as a fundamental actor in bringing about this future, and the nation therefore assumes a global role by casting its collective eye forward toward an imagined future in which it remains or becomes significant. We can situate the *mission civilisatrice* assumed by France, like the United States' hero complex, squarely within this anticipatory aspect of the nationalist geopolitical vision. Third, geopolitical motifs involve a particular conception of national boundaries, core areas, and the relative importance of various areas within the national territory. This territorial aspect may defend existing state borders (e.g. France) or claim areas outside existing borders (e.g. Armenia with regard to Nagorno Karabakh). Yet again it may demand a redrawing of national borders (e.g. Kurdish and Basque nationalism). Thus it is partly defensive and partly antagonistic. It also upholds certain foreign nations or states as examples (good or bad) to serve as guides to action. Fourth, new events interpreted through the lenses provided by past experiences are greeted with the *habitual attitude* of a particular nationalism, such as curiosity, fear, defiance, or opportunism, to list only a few of the more common nationalist attitudes. Fifth, in conjunction with these temporal, spatial, and attitudinal factors, geopolitical motifs promote certain grand ideologies: theories and principles, belief systems, and philosophies. Examples of such ideological elements include capitalism, democracy, and a religion (most often Christianity, Islam, or Buddhism), as well as more specifically geographical notions such as "the mysterious unity of *Mitteleuropa*, and the domino theory of communist expansion" (Dijkink 1996, 14).

To sum up the points introduced thus far in simpler terms, every geopolitical motif (representation) involves a geographical theory (conception) situating the nation in regard to its neighbors and the rest of the world, as well as historical and ideological frameworks (related conceptions) giving meaning to territories belonging to the past and the future, to "Us" and "Them." Through these various spatial, temporal and ideological facets of the geopolitical vision the nation and its Others are given meaning, and this meaning in turn is a source of national power and personal and group power within the nation.

Yet speaking of national power raises complex issues, not least of which is that, as we have seen, the medium through which one encounters an event of global importance (perception) can determine which geopolitical conceptions are activated because the peculiar affordances of a medium constrain and enable representational opportunities. In addition, representations of nations evolve or fail to evolve while nations themselves undergo transformation.

> A world role is linked to identity. At the beginning of the twenty-first century all nation states, particularly Western ones, are questioning their national identities. States with pretensions to an international identity and world role are having to do so more acutely. World powers had identities that the international community could recognize. Britain was a world empire promoting notions of self-government, free trade and the rule of law; the USA was the champion of liberal democracy and freedom; … France was the foyer of the universalist values of the Rights of Man, the French language and culture. But today those personalities have disappeared, are contested or have withered.
>
> (Keiger 2001, 232)

This quote points to the poignancy of the current moment in national history and in the history of national representations. It is a time in which great powers may derive strength more from the opinions of foreign observers than from their ability to inflict economic and military damage on opponents; reputation is everything.

Soft Power and Global Democracy

The risks involved in permitting the national reputation to deteriorate can be understood in terms of a distinction between "hard power" and "soft power" (Nye 2002, 2004). Hard power allows a country to obtain what it wants despite resistance, and it is evident in military muscle and economic supremacy. Complementing hard power is soft power, the ability "to entice and attract," to obtain desired ends by persuading other countries to take a certain course or by having interests that correspond to courses of action that are inherently appealing (Nye 2002, 9). In the arena of soft power national reputation is everything; an attractive culture, contagious ideology and admirable institutions work together. The importance of soft power in maintaining social order is increasing with globalization and the diffusion of new communication technologies. Joseph Nye frames this situation in terms of "nuggets of hate [that] are unlikely to catalyze broader hatred unless we abandon our values and pursue arrogant and overbearing policies that let the extremists appeal to the majority in the middle" (Nye 2002, xi).

Hard power is not sufficient to deal with new global problems such as international drug trade, AIDS and other epidemic diseases, ecological disturbance, and terrorism since they must be handled with international cooperation, which is produced by soft power: "If a country can make its power legitimate in the eyes of others, it will encounter less resistance to its wishes." (Nye 2002, 10). Hard power is of course essential for ensuring national security, but it cannot stand alone, and in combination with unilateral policies and arrogant rhetoric it can actually stimulate the emergence of coalitions of opposing states (Nye 2002, 39). This is why France's role as "self-proclaimed champion" of cultural particularism is not as ridiculous as John Keiger makes it out to be (2001, 225). Such a symbolic stance can be coldly appraised as an effective defense strategy.

Nye makes a major contribution by bringing this situation to light, but what Nye does not emphasize is that the formation of concerted opposition to the US and its hard power is unlikely to simply check US power. Radical political changes are often accompanied by violence, particularly when a former power is overthrown, as in the French Revolution, Russian Civil War, and the Chinese Cultural Revolution. These upsets took place at the national scale but future upsets of a similar kind will sooner or later transcend the national scale.

Reliance on hard power does not simply threaten US supremacy, in the long run, but also generates tension that could be released catastrophically. The US's role is unprecedented and its supremacy may itself be an inevitable threat to democracy. Nye therefore fails to recognize that the concern with soft power cannot be justified in terms of ensuring US supremacy, even a "soft" supremacy. Soft power is in fact better understood as an indirect effect of general respect for transborder democratic processes which in the long run will confer their primary benefits not on any single state but on the citizens of many states, including Americans.

Conclusion

This chapter has examined geopolitical representation in the abstract, linking representation to perception and conception. In this context we can speak of geopolitical "motifs" and analyze their functions. The motif of "America" appears as a kind of symbolic glue that holds together otherwise disparate worldviews, and the prevalence of America in geopolitical discourses around the world is not explicable simply in light of the strength of the US, but more subtly it reflects the usefulness of a powerful competing state as a symbolic motif (good and bad) within various local political contexts. Because representations of states from other states change quickly from positive to negative or vice versa depending on economic, political, and cultural conditions, the conditions we observe today may change rapidly. For this reason and others, attention must be given to geopolitical representations and their impacts. We are justified in adopting Nye's distinction between hard and soft power but must take care not to assume that soft power serves only the needs of the state in question. By cultivating soft power in a system of democracies, a state will inevitably contribute to the evolution of the global public sphere and towards an unbounded form of democracy.

The internal and external geography of the nation play important roles in geopolitical perceptions, conceptions and representations. Each nation's geopolitical context shapes its rhetoric (representations), understanding (conceptions) and worldviews (perceptions) and this triad in turn shapes its evolving geopolitical situation. So in the case of France, the America that is *seen* is shaped by the France that has *been* – that is, the historical geography of France including key events in this country's emergence as an important European state, and its forging of a "certain idea" (as de Gaulle would say) of the nation. Also important is the imagined future associated with French ideas of the nation and national identity, particularly fears of continuing decline in the face of relentless competition from American culture.

While voting remains tied to territorial entities – states and their components – the rest of the democratic process has expanded in many ways into Appadurai's "ideoscapes" or Luke and O'Tuathail's "flowmations." These structured flows of international communications depend on a wide (and expanding) range of media which vary in their affordances, presenting different political possibilities and constraints. Tracing this phenomenon is the first step towards addressing the failure of the state (the US or any state) to produce soft power commensurate with its hard power. By avoiding the territorial trap and theorizing democracy in an unbounded way we not only can begin to consider the transnational circuits of nationalistic motifs, but we can also recognize the gradual replacement of states by larger, supranational organizations like the EU. The significance of the study of representions of a nation therefore lies in the connections that can be traced between particular geopolitical motifs and corresponding ways of perceiving/seeing/envisioning as well as conceiving/understanding/interpreting.

Chapter 3

France-US Relations and the 2004 Election

> War does not always give democratic societies over to military government, but it must invariably and immeasurably increase the powers of civil government; it must almost automatically concentrate the direction of all men and the control of all things in the hands of the government. If that does not lead to despotism by sudden violence, it leads men gently in that direction by their habits.
>
> de Tocqueville, *Democracy in America*, p.650

In 2004, American conservatives sought to "contaminate" John Kerry with the tainted image of France. France was seen as the natural enemy of the US since it had opposed US plans to invade Iraq. The reasons underlying French opposition to the American invasion of Iraq were disregarded and France's support of the Afghanistan invasion only two years earlier was forgotten. At the same time, John Kerry, as the Democratic nominee for President, was the main domestic opponent of the Republican Party. France was therefore linked to John Kerry and perverse creative energy went into reinforcing this symbolic association in public perception.

Digital images were crafted and posted on websites showing a Frenchified Kerry. One digital editing gave him a beret and a "United Nations Uber Alles" pin and placed his portrait under the heading "John French Kerry". Another image again gave him a beret, as well as the word bubble, "French, Moi?" while a third superimposed his face on a French flag and flanked him with images of the Eiffel Tower and a frog. A bumper sticker in a similar vein bore the image of a jubilant Jacques Chirac and the inscription "Make the French Happy! Vote for John Kerry" (Fig. 3.1). These images reached not only their intended audience of nationalistic American conservatives but they also broke through the thin and permeable boundary between political communities. On March 26, 2004, Pascal Riché, a reporter for the left wing French newspaper *Libération*, posted copies of anti-Kerry and anti-France images on his "blog," and even offered his French audiences a link to Rush Limbaugh's website where one could find a reference to "Jean Chéri."

The strategic construction of a Kerry-France association was a prominent facet of the resurgent nationalism of American conservatives in the post 9–11 period. This nationalism circulated within the Bush administration, the Republican Party, and the Right in general as a response to the perceived threat of cosmopolitan and internationalist values which had gained power during the Clinton administration. Bush was portrayed as reasserting global order and democracy through American

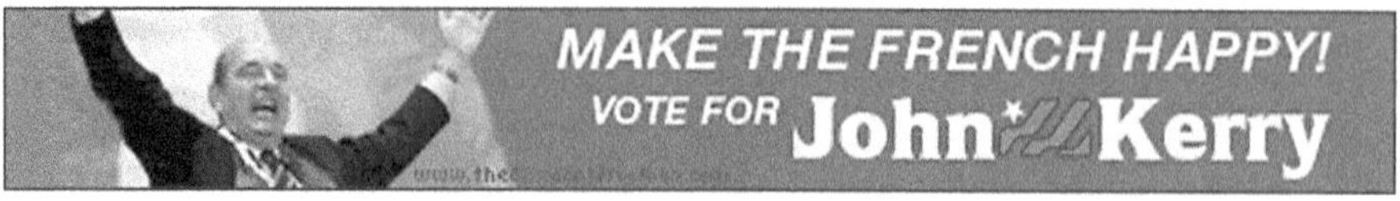

Figure 3.1 Bumper sticker design mocking France and John Kerry. Source: courtesy of http://www.thedissidentfrogman.com, 2004 also posted on http://www.liberation.fr.

"leadership," meaning withdrawal from negotiations on several international treaties including the Kyoto Protocol, the Anti-Ballistic Missile Treaty, and the International Criminal Court, an approach rationalized by right leaning intellectuals of statecraft such as Gary Dempsey of the Cato Institute: "True leadership means pursuing policies that are in America's national interest, and persuading other countries that the policies are in their national interest too. It does not mean, as some of the president's critics contend, doing things because they will make other countries happy" (Dempsey 2002). While Kerry had sterling military credentials he could be attacked on the basis of his European ties, his diplomatic concept of foreign relations, and most insidiously for his failure to match a certain vision of American masculinity (California Governor Arnold Schwartzenegger launched this attack with the term "girlie-man").

Xenophobia, partisan politics, and nationalistic delusions of grandeur therefore linked Kerry to France. Conversely, appeals to internationalism, cosmopolitanism, and tolerance were rejected, as they have been by nationalists in many other times and places, with a mix of moralistic condemnation and misinformation (Dodds and Atkinson 2000). It was this rhetorical climate that shaped representations of John Kerry in 2004 in both positive and negative ways. The Democrats felt a need to play up their candidate's military credentials to the point of absurdity in the attempt to neutralize the equation of their party with weakness and unpatriotic compromise, while the Republicans succeeded in tainting Kerry with an aura of foreignness that blended racial, sexual, linguistic and religious slurs and demonstrated the irrelevance of "political correctness" to the majority of the American public.

Francoise Meltzer, Professor of Romance Languages at the University of Chicago, articulated this tactical maneuver: "Not only does Kerry speak French. He speaks English well. In addition his wife is foreign – she speaks with an accent and she speaks her mind… That further contaminates Kerry. He's part Jewish, he grew up Catholic, he studied in Switzerland and he speaks French – this all combines to make him 'French,' not really American" (Maler 2004). The symbolic contamination of Kerry was based, in short, on a complicated us/them opposition that transformed a war on terrorism into a dictate to vote against the "French" candidate. Generally this device was employed by the Republican party and conservative stalwarts rather than by the President himself, but he did declare pointedly in a campaign speech at Allentown, Pennsylvania: "The use of troops to defend America must never be subject to a veto by countries like France" (Bush, 2004). This rhetorical device caught the attention of the French government, which responded through its ambassador, Jean-David Levitte, who called Dan Fried of the National Security Council and "explained with all necessary tact that the Republican candidate could no doubt attain his electoral objectives without mixing France up in the campaign..." (Le Figaro 2004a).[1] In theory, this was true but France was a powerful political tool for conservatives no longer able to solidify domestic support by brandishing terms associated with skin color or the defunct communist threat.

Taking these observations into account, France's importance for the study of geopolitical representations of America in 2004 derives from three distinct sources: the symbolic association of Kerry with France by his opponents, the use of France as a gendered and othered object of derision by conservatives, and the support for John Kerry in French popular and intellectual discourses. This chapter exposes the wave of French-bashing that struck the US during the George W. Bush administration then moves on to consider French views of the US and of the candidates. This leads to a reflection on US-France relations in general then a consideration of France-Europe relations. Most intriguingly, we will see the dependence of Franco-European relations on Franco-American relations, and vice versa.

France-Bashing: High-brow and Low-brow

American talk show hosts took advantage of the chance to mock the French, playing on tired themes of cowardice. Jay Leno and Conan O'Brien (NBC), Dennis Miller (CNBC) and David Letterman (CBS) all joined in the chorus. Leno quipped: "Did you see the new bomb the government came up with? It weighs 21,000 pounds. The Air Force tested this bomb in Florida and the bomb blast was so strong at Disneyworld 25 French tourists surrendered." On a similar theme, he joked that: "French troops arrived in Afghanistan last week, and not a minute too soon. The French are acting as advisers to the Taliban, to teach them how to surrender properly." France has become the joke target *par excellence.* The Subway fast food chain, a multinational based in Florida, promoted their cordon bleu chicken sandwich with the words "France and chicken, somehow it just goes together" and a photo of a chicken dressed like Napoleon (Agence France Presse 2005). A political commentator on Fox network observed that the Olympic Committee had "missed a golden opportunity" in not choosing Paris for the 2012 Olympics, because the terrorists could "blow up Paris, and who cares?" (Agence France Presse 2005). If talk show hosts are known for their tendency to mock virtually everything, the same is not true of advertisements and the news. Likewise, if talk shows and the news regularly touch on political and geopolitical motifs, ads do not normally do so. By making a joke of France, aggressive nationalism was normalized and became acceptable in an increasing range of situations. The incorporation of France-bashing in all these contexts in the 2004–2005 period therefore indicates normalization of the anti-France discourse in American culture, a trend that diffused from, and fed into, an expansionary geopolitical agenda of the radical Right.

1 All English translations in the text are by the author unless otherwise noted.

The question this discursive event raises is why it was so important to dismiss the *right* of France to disagree with the US and to transform its show of resistance into cowardice, hysteria, duplicity and irrationality? Something special was involved in the symbolic treatment of France, as indicated by Condoleeza Rice's comment (made to her associates in spring of 2003): "punish France, ignore Germany and forgive Russia" (Leicester 2004). France was not simply an opponent to the US policy in Iraq, it was a *useful opponent.* By denigrating French concerns, French politicians, and French media, the pro-war segments of US society and government could strengthen their position. France was, quite simply, a scapegoat. Ironically, as we will see, American nationalism was useful for many politicians in France as a means of building political cohesion at the scales of both France and Europe.

Schwartzenegger's comment directs attention to the gendered aspect of Atlantic reverberation. Geopolitical motifs are often if not always gendered, and femininity adheres to few national images in the American imaginary to the extent that it adheres to France. The country is seen as pretty, romantic, impractical, and helpless (the last trait being "proven" by several wars). The dynamic includes not just the nation but also its people. The French woman is idealized by Americans as the paragon of femininity while the French man is lampooned for excelling at frivolous, sensual, "womanly" things like painting, philosophy and cooking. As the US expands its global military and economic control, motifs of sexual dependence like Arnold Schwartzenegger's "girlie-men" become more prevalent as a means to "explain" international relations. These gendered stereotypes and their reflection in national relations have been accentuated by a president who votes and vacations in Crawford, Texas (playing the role of the quintessential American manly man, the cowboy) and a resurgence of homophobia and sexism in grassroots politics.

As US-France tensions in 2004 created opportunities for all kinds of communications, books that stoked the fires of xenophobia, nationalism and chauvinism were easy to sell. Kenneth Timmerman's *The French Betrayal of America* (2004) serves as an example. The dust jacket of the book shows the face of the Statue of Liberty with a tear dripping from her eye and the pages carry an account of US-France relations filled with outrage that France insists on conducting an autonomous foreign policy. France's disagreement on topics such as the war with Iraq demonstrates, on this account, an outrageous lack of gratitude because the country owes its very existence to the US and has therefore permanently forfeited the right to exclude American troops from French soil, to question American policy or to disagree with the US. Furthermore, by helping to arm Iraq (albeit during a period when the US was also arming Iraq) France has demonstrated its profound untrustworthiness. Thus although some French citizens may be worthy of respect as individuals, the French nation should no longer be favored with scientific or military cooperation by the US.

Very similar in intent, though more contemptuous than outraged, is Denis Boyles' *Vile France* with the smirking subtitle: *Fear, Duplicity, Cowardice and Cheese* (2005). In Boyles' view, "France is a rogue nation in the heart of Europe, and therefore one perhaps even more dangerous to the world than North Korea because we in the United States fail to take its animosity and recklessness seriously" (2005, 78). His scorn mounts to the final pages in which he writes: "Yet only a French foreign minister would cap a couple of years of sabotaging American foreign policy, running up a tab that can be calculated in American lives and treasure and undermining international institutions by writing an open letter to America lecturing and scolding us about our behavior toward his underhanded, duplicitous, insignificant little country with its pretty but largely useless language" (Boyles 2005, 160).

A less entertaining but only slightly less biased analysis of US-France relations can be found in Jacquelyn K. Davis' *Reluctant Allies & Competitive Partners* (2003). Published under the auspices of the Institute for Foreign Policy Analysis (IFPA), a conservative think tank that bills itself as "nonpartisan," [2] this book relentlessly depicts French foreign policy stances as weak, self-serving, irrational and hypocritical. Like Timmerman, Davis interprets the French refusal to accept US arguments for invading Iraq in 2003 as a sign of ingratitude: France's obstinate refusal to believe American claims regarding Iraqi weapons of mass destruction and Saddam's links to al Qaeda shows nothing less than France's failure to "appreciate the sacrifices that Americans have made in the name of European security" (Davis 2003, vi). The language of the book alternates between open contempt for the French and subtle innuendo designed to make French leaders appear underhanded and deceitful. Thus, the French engaged in "transatlantic sniping" after September 11 and it was "Gallic logic" that led them to believe the US was setting a bad precedent by disregarding international law (Davis 2003, 42, 153). "French leaders should be aware that their behavior is alienating more and more of their allies, leading them to seek ways to work around the French when faced with their intransigence" (Davis 2003, vii) although the same could be said with more assurance about the US. Conservative fervor in the US spurred the sales of books like those of Timmerman, Boyles and Davis putting France "in its place" geopolitically speaking.

The moral outrage of American nationalists inevitably turns against the French media. I will quote at length a "Letter from Paris" by Joseph Harriss, printed in the December 2004/January 2005 issue of *The American Spectator*.

> Sensitive, artistic, intelligent, nay, *intellectual*, souls that they are, the French have never been known for steady nerves and grace under pressure. Thus the collective *dépression nerveuse* (freely translated as a hissy fit) they have been going through since the American people resoundingly re-elected George W. Bush. The deep trauma they have suffered is directly proportional to the high hopes they had for John Kerry.

---

2 The Institute for Foreign Policy Analysis enjoys financial support from conservative foundations such as the Carthage Foundation, the Lynde and Harry Bradley Foundation, the Sarah Scaife Foundation, as well as military contractors such as Raytheon, Rockwell International, McDonnell Douglas, Westinghouse, G.T.E. and Boeing Aerospace (www.sourcewatch.org, a website of the Center for Media and Democracy).

> As one man, the French media, that good and faithful servant of the Élysée Palace and Quai d'Orsay, painted an adoring portrait of Kerry as Someone Like Us. He spoke French, it was repeatedly pointed out, spent his childhood summers in France, had a French cousin, and, *mon dieu!*, a wife who could say hello in five languages. He understood Us. So unlike that boorish, cretinous, bellicose religious fanatic (a Texan, of course) who had been elected by accident, or probably fraud. In Germany the bestselling newspaper, *Bild*, might have good words for Bush, but *Le Monde*, the thinking Frenchman's newspaper, couldn't resist sticking its nose in America's business with an editorial calling for Kerry's victory.
>
> (Harriss 2004–2005: 64)

In Harriss' worldview what makes the French contemptible, weak and pathetic is that they fail to accept US domination. If there is an element of paradox in this view, it is not one that nationalists like Harriss are likely to perceive. While his essay makes no pretense to objectivity, its bitterness suggests an affront, and would be difficult to explain unless there was in fact an anti-American bias in the French media. Accusations of media bias are confirmed by the well-known specialist in European history, John Keiger who judges that in France: "Remarkable press and media consensus reigns over the evil of US-led 'ultra-liberalism'" (2001, 223).

Yet the situation is more complex than it appears. Rather than an overt bias against Bush and open support of Kerry, the French media (with the exception of the daily *Libération*) in fact critiqued various elements of American society and various regional cultures that supported Bush, while at the same time leaving unresolved the question of which candidate would be better, whether Bush actually represented American opinion in general, and whether France would benefit from a Kerry victory. That is the story taken up in the following chapters. Here we will turn to a final example of academic France-bashing.

Robert Kagan, former Secretary of State under Ronald Reagan, published in 2002 an explanation of why the Europeans failed to properly appreciate the United States in *Policy Review* (Kagan 2002). Translated excerpts quickly appeared in *Le Monde* (July 27 and 28, 2002) and somewhat later in the French scholarly journal *Commentaire* (no.99, winter 2002–2003). What is particularly striking about Kagan's argument briefly stated above, is its assumption of a link between political culture and military power. He argues, in essence, that military power and weakness encourage two different kinds of political culture and two different political philosophies, one clear-headed, the other deluded. The realist, "Hobbesian" approach to geopolitics is evident to inhabitants of powerful states, who must respond to real-world conditions, while the idealist, "Kantian" approach is attractive to those living in weak states, whose protected condition fosters the illusion that international laws should be respected. If we strip away the technical language Kagan is arguing, in essence, that pacifist and diplomatic Europe embodies the naïve worldview of a sheltered child while the militant and uncompromising US embodies the mature worldview of an adult. Or in a more sexist language: Europe is the woman with her charming attachment to manners and etiquette while the US is the man – coarse and rude but good to the core and willing to risk his life and break a few pointless rules to protect his beloved family from misfortune, thieves and vagabonds. This paternalistic view is rooted in American images of the noble outlaw, "a destroying angel with spurs who protects decent farmers and ranchers from exploitation by tyrannical figures of authority" (Tatum 1982, 111) while the tyrant merges here with the terrorist. Sexism runs through Harriss' critique as well, since a "hissy fit" is a feminized form of disagreement implying irrationality and hysteria, while tolerating but resisting such emotional performances is the mark of a real man.

Kagan's article provoked a sharp response from French intelligentsia who questioned whether the US's geopolitical realism was actually realistic. They asserted that Europe's Kantian idealism was in fact the most practical way to respond to the tensions arising from globalization while taking account of globalization's constraints and opportunities. Their critique did not reverberate back to the US because of the politics shaping the asymmetrical flows making up Atlantic reverberation.

From bumper stickers to political science, American discourses of 2003 and 2004 seized on France as a symbolic target. From anti-France jokes and advertisements to slanted attempts at scholarly critique Americans took aim at what was only partly an external enemy: the terrorist in the hi-jacked jet merged with the skeptical ally in Europe and the Democratic challenger on the campaign trail. Despite such political fusions, the general public in the US remained ignorant of the real friction between France and the US at the official level – from gestures such as Jacques Chirac's refusal to stroll on the beach on Sea Island, Georgia, with the other G-8 leaders to Vice President Dick Cheney's studious avoidance of the French ambassador living across the street from his house – and few Americans understood that the stakes of the struggle were the loyalties of smaller and "newer" European countries.

Taking Sides

To what extent was the American Right's conflation of France and Kerry reflected by a real preference for Kerry in France? Statistics help shed light on this question. International survey results presented in *Le Monde* showed that among 10 countries – France, South Korea, Canada, Spain, Mexico, Australia, Japan, United Kingdom, Russia, Israel – Kerry was preferred in all countries except Russia and Israel, and was preferred by the highest percentage of the population in France (Table 3.1). Likewise, France registered extremely weak support for Bush (with only Spain posting a lower score on this measure). What is most telling is the ratio of French respondents preferring Kerry to Bush, which was also highest in France at 4 ½ to 1 (Table 3.2). In siding with the challenger, the French broke from their customary support for the incumbent American president (Barochez 2004).

An in-depth survey conducted in France immediately following the election indicated that dislike for the incumbent President was slightly stronger among men (67%) than women (63%), much stronger among persons aged 15–34 (> 80%) than among older cohorts (< 65%), and stronger among mid-level professionals and employees (> 79%) than among those of lower *or* higher socioeconomic status

Table 3.1 Preferences for George Bush and John Kerry: by country, fall 2004. Source: *Le Monde* 16 October 2004, p.18.

Country	Percent that prefer Kerry	Percent that prefer Bush
France	**72**	16
South Korea	68	18
Canada	60	20
Spain	58	**13**
Mexico	55	20
Australia	54	28
Japan	51	30
United Kingdom	50	22
Russia	48	52
Israel	24	50

Table 3.2 Four and a half times as many French respondents to this survey supported Kerry as supported Bush, a degree of pro-Kerry preference matched only by the Spanish. Source: *Le Monde* 2004, p.18.

Country	Ratio of Kerry supporters to Bush supporters
France	4.50
Spain	4.46
South Korea	3.78
Canada	3.00
Mexico	2.75
United Kingdom	2.27
Australia	1.93
Japan	1.70
Russia	0.92
Israel	0.48

(laborers 48%, the unemployed 59%, top-ranked professionals 72%, and managers 74%) (CSA 2004b).[3] As for interest in the election, a survey just prior to the election indicated that half of the French population was interested in the election, with stronger interest from men (51%) than from women (40%), stronger interest from the 18–24 year-old cohort (62%) than from older persons (> 46%), and stronger interest from managers and professionals (> 65%) than from mid-level professionals (44%), manual laborers (33%), farmers (46%), or wage laborers (36%) (CSA 2004a).[4] In short, the most interested French observers were young professional men, precisely the French group that most strongly favored Kerry over Bush. Thus, a strong opinion about the outcome of the US election was present in France, and the level of favoritism was greater than that in any other surveyed country. Furthermore, a key trend-setting group in society – young male professionals – manifested the strongest anti-Bush opinion within French society. On the question "In your opinion, would George W. Bush or John Kerry be more capable of improving Franco-American relations" more than 7 times as many respondents picked John Kerry, and on a similar question related to US-Europe relations the ratio was equally high. The French also placed much greater faith in Kerry than in Bush to protect the environment, stabilize Iraq, and improve the American economy. They were least pro-Kerry when asked who would be better at fighting terrorism, nevertheless Kerry was still favored approximately 2:1 (CSA 2004a). As we will see, this opinion was not a result of indoctrination by the French media, but actually deviated from the cautiously neutral focus and perspective of the media.

What is not clear, however, is the degree to which the survey numbers indicate genuine approval for Kerry or alternatively an "anyone but Bush" stance. The latter would not be surprising given the French concept of leadership which draws on royalist and republican traditions to construct the ideal leader as imperious or intellectual, but in either case remote and somewhat alien, whereas the ideal of an American leader is someone whose interests and mannerisms are markedly ordinary. The French saw more of their own brand of politician in Kerry while Bush appeared not only to be quite mundane, but to actually revel in being mundane, to flaunt his truncated vocabulary and his lack of knowledge about the world – not at all the way a French president would behave. To what degree did the French judge Bush, the man, and to what degree did they see him through a lens of America personified? To what degree was the reaction to Bush an expression of anti-Americanism? How much of the French preference was orchestrated by French media bias?

---

3 Survey specifications: "Sondage exclusif CSA / LE PARISIEN / AUJOURD'HUI EN FRANCE réalisé par téléphone le 3 novembre 2004. Echantillon national représentatif de 780 personnes âgées de 15 ans et plus, constitué d'après la méthode des quotas (sexe, âge, profession du chef de ménage), après stratification par région et catégorie d'agglomération."

4 Survey specifications: "Sondage exclusif CSA / FRANCE EUROPE EXPRESS / FRANCE INFO réalisé par téléphone les 27 et 28 octobre 2004. Echantillon national représentatif de 1000 personnes âgées de 15 ans et plus, constitué d'après la méthode des quotas (sexe, âge, profession du chef de ménage), après stratification par région et catégorie d'agglomération."

The numbers alone cannot shed light on these questions, nor can they tell us how French citizens arrived at their preferences, or exactly what kind of negative and positive feelings they nurtured. Fear of America and dislike of Bush might both produce a similar preference for Kerry. For this reason, it is essential to engage with discourses that reflect sentiments as they circulate through media and society. This attempt to extract geopolitical meanings from media texts will be the focus in the chapters that follow. First, however, we must take the time to consider Franco-American relations from a broader perspective.

French Critique and the Achilles Heel

> *Toujours Paris s'écrie et gronde*
> *Nul ne sait, question profonde*
> *Ce que perdrait le bruit du Monde*
> *Le jour où Paris se tairait.*[5]
> (Victor Hugo)

The reverberation of snubs and rebuffs does little for a dominant state but it does much for a weaker state in terms of global opinion. In other words, among an international audience Védrine's critique of the American hyperpower has more rhetorical leverage than the American France-bashing, academic or otherwise. This is because of the natural tendency to sympathize with the weak and rebel against the powerful, particularly, it should be noted, when one is socialized into a democratic society. This is one example of what I call the "Achilles heel" of a hyperpower or superpower. Symbolic attacks, everything else being equal, can best serve the interests of their authors if directed at the most powerful contender when democracy's *egalitarian ethos* is a central part of the global communication environment. Soft power, like water, tends to flow downhill. A trade of verbal barbs between intellectuals of a superpower (or hyperpower) and those of a weak opponent or competitor is more likely to rally support behind the small power, as long as the verbal barbs are not accompanied by violence (a tactical failure on the part of the weak because it undermines their claim to innocence). Védrine argues that the term *l'hyperpuissance* indicates the simple fact that the US is essentially unchallenged whereas the term "superpower" was applied to global powers that faced off against other comparable powers. A *hyperpuissance* is a giant with an Achilles heel in the form of its unrivaled *utility* as a symbolic target. While it is true that in French the prefix hyper- lacks the negative connotations carried by the same prefix in English, in the context of French worldviews, and most democratic nations' worldviews, the "hyperpower" image cannot be evoked without provoking some measure of anxiety and unease. If the sovereignty of the people is compromised by any element of globalization, blame is

5 "Always Paris yells and scolds/No one knows, profound question/What would be lost from the noise of the world/if Paris should one day hold its tongue." (author's translation)

naturally directed at the most powerful outside power. A hyperpower that is far from blameless simply invites rhetorical attack and the subsequent loss of legitimacy.

Domestic struggles outside the US are likely to be shaped for some time by the "mileage" political figures outside the US can obtain by taking anti-US positions or drawing attention to US military, economic and cultural dominance. The rhetorical power conferred on French communications by the use of America as a motif draws on notions of citizenship and self-determination that are familiar to Americans in a domestic context but that seem entirely new when directed at the US from abroad. The dynamic also assists France in their leadership position in the EU, in a way that has already and will continue to challenge the US. Of course not all critiques of the US are primarily attempts to exploit the "Achilles heel" but the severity and duration of critiques of the US are now greatly increased by the hyperpower's preponderance of hard power and its growing dependence on this approach.

To emphasize the concept of hard power as a liability we can note that the percent of French citizens with a favorable opinion of the US has fallen from 63% to 31% since 2002 (Pew 2003) precisely during the time the US has demonstrated its political and military power relative to the UN and Iraq. Of course France has a history of concern and skepticism regarding American culture and the US's global role, but the corresponding motifs of critique are activated by the policies of the Bush administration that unfortunately fit quite well with the anti-American stereotypes that have been held for centuries by the French (Roger 2002) and other Europeans.

Although the French trace their origins as a people into the distant past of the Capetians, Gauls and Romans, France's origins as a republic are very close in time to the origins of the United States. Both republics laid claim to democracy as a national invention in the late 18th century, and both countries continue to promote the idea of being the standard bearer for democracy itself. Other similarities can be found: both are wealthy countries; both are among the world's five largest economies; both hold permanent seats on the UN Security Council; both were among the first five countries to develop nuclear weapons and the first five nuclear powers recognized under the Nuclear Nonproliferation Treaty; Paris, like New York, ranks among the four "world cities" that drive various facets of globalization (Beaverstock, et al., 1999). In addition, French and American nationalism are both nourished by the idea of being special – historically and geographically *exceptional.*

Most important to this motif of exceptionalism is that both countries have been perceived over their respective histories, by leaders and citizens alike, as having a global *mission.* And this mission, in both cases, is supposedly to spread civilization to the other peoples of the world. When Americans look at France they become "perplexed over whether the phenomenon of a country that must possess grandeur if it is to be itself is inspiring or simply ridiculous" (Bernstein 1990, 17). But they fail to see their own indoctrination with the identity of "world leader." Perhaps as much as the ideal of spreading civilization itself, it is the *talk* about this mission that exasperates Americans: "more down-to-earth, less idealistic peoples, such as the British, the Dutch, the Germans, conquered with a good deal less fancy talk, simply to enhance their wealth and power" (Bernstein 1990, 15). Likewise those Americans most invested in their own country's civilizing mission are least favorably inclined toward philosophical musings about the fine points of what civilization entails.

In France, the expansive geo-historical role is referred to as the *mission civilisatrice* while in the US (where only in the late 20th century has it replaced isolationism) it goes by a host of titles such as "spreading freedom," "promoting democracy," "protecting liberty" and the like, and it builds on the 19th century concept of manifest destiny embodied in the vision of American territorial conquest as God's will. This pervasive worldview was manifested, for example, in the Cold War articles of *Reader's Digest* in which "Stories about communism and the 'Soviet way of life' were counterposed with representations of 'America' that tended to emphasize not only national territory but also a transcendental morality, an idea which meant the United States had a manifest duty to protect democracy and capitalism for the wider world" (Dodds 2000, 83; Sharp 2000).

Such extensive similarities between France and the US do not preclude disagreement about the *means* of supporting civilized attitudes or about the role of various countries in the world system. Indeed, there cannot be two global leaders on an equal footing, so the US views the leadership claims of France and Europe with bemusement while the French view the US claim with a mixture of scorn, dread, and wounded pride.

In addition, the similarities in motifs regarding the national geo-historical role do not mean cultural and philosophical differences are easily overlooked. The differences are all the more troublesome precisely because of the similarities. Brenner and Parmentier argue that "of all the long-standing connections between allies in the Western world, the French-American relationship is undoubtedly the most unsteady" (2002, 2). French historian Ran Halevi characterizes the Franco-American relationship as a peaceful disagreement (*mésintelligence pacifique*) which "appears devoid of collective attributes that would produce lasting affinities or hostilities" (2004, 27), while John Keiger documents the profound ambivalence of the French towards the US – seeing a "sister nation" and a long-term ally, yet also an alien culture driven by acquisitiveness and Protestant values, instrumental in the worldwide decline of the status of the French language. Like Nye (2002, 15), most Americans adopt the *hegemonic stability thesis* and believe that "global governance requires a large state to take the lead" (see Dodds 2000, 21). In contrast, most of the French would prefer a world without a strong leading state. The tension is exacerbated because of a chasm between the communication patterns of the French and Americans. Their languages are grammatically and phonemically different and they have sharply differing attitudes toward argumentation, debate and criticism (Carroll 1988). Misunderstanding proliferates in the cracks between these divergent communication habits even when one or both communicators are bilingual. Meanwhile, the intensity of trade and frequency of contact between representatives of the two countries (in forums such as the UN Security Council, NATO, and the G-8) cause recurring irritation to diplomats and bring their cultural, political, ideological and communicational differences to the fore. With the foundering of the Soviet Union and the terrorist attacks of September 11 the tension has mounted with little relief. Observers on both side of the Atlantic have accordingly characterized the current phase of Franco-American relations as one of the chillier phases of the cold alliance.

World War II initiated the modern period of Franco-American relations. In this phase, Americans expected the French to recognize that they were indebted to *and therefore* dependent on and subordinate to the US. But the French did not always do so, and rarely to the degree that the Americans would have wished. Whatever its justifications, the American refusal to treat de Gaulle and *France Libre* as viable military resources caused lasting bitterness. It did not help that de Gaulle's return to France was strategically delayed and Europe was restructured at Yalta without his participation. These snubs left serious scars including the separation of France from NATO's integrated military command in 1966, creating a third nuclear bloc in the Cold War theater and a fundamental break in geopolitical space that has endured long after de Gaulle's disappearance from the French political scene. "France has pursued its interests by making it too painful to ignore them" (Kissinger 2001, 50, quoted in Davis, 2003, 49) creating tension with the US over issues such as the development of nuclear weapons, the Middle East, and most recently during the 2003 invasion of Iraq. France has also taken the lead in supporting European military coordination separate from NATO under the auspices of the European Security and Defense Policy (ESDP).

The Gaullist position on foreign relations is summarized by Henri Froment-Meurice (2004–2005, 951): "the real way to counter America's permanent temptation to lead 'the West' and the 'free world' was to provide balance with a politically, diplomatically and militarily strong Europe." While the US looks back to the beaches of Normandy and is surprised at France's coolness, France looks ahead to the world of the 21st century and is wary of the US's protective embrace. "For General de Gaulle, rapprochement with Germany had the political goal of creating an alternative force to the United States, or at least demonstrating independence from the US. For him, a grouping around France and Germany was a necessary condition of the emergence of an economic and political power capable of rivaling the United States" (Cohn-Bendit 2005, 81). Germany welcomed the French overtures, but as part of a general opening up towards the west, not in the spirit of creating an alternative to US power. While courting German cooperation, the French were willing to overlook this *mésentente*. Economic interest kept them on board even if "The common market, according to de Gaulle, was to be an organization whose cooperation and policies would be inspired by a French vision – namely, that of a third great power bloc in the world buffering the Soviet-American antagonism" (Bernstein 1990, 317–18).

France's unique position of non-alignment among the Cold War powers meant that "the misunderstanding between de Gaulle and Roosevelt, even if explicable through 'character' and contingencies, signaled from the Second World War onwards the real difficulty experienced by these countries of the oldest alliance that had never been at war, in really making peace together" (Froment-Meurice 2004–2005: 951). This was a rivalry marked by a new form of international power – hegemony. Henry Luce's "American century" (Baughman 1987, 129–37) spelled the end to the colonial era in which European powers had carved up the world and the beginning of a new way of dominating and controlling other countries. The Suez Crisis of 1956 was a crucial signal of the waning of empire and the rise of hegemony. It was harder for the French than for the British to accept this intrusion of American strategic objectives in what had seemed for more than a century like European territory (Keiger 2001, 180). The difference arose in part from the cultural and linguistic similarities between the US and the UK, whereas both language and culture divide the Americans from the French.

As Philippe Roger (2002, 392) argues, the positive feelings generated by the Franco-American alliance during wartime have been asphyxiated repeatedly by disagreements surging in the interwar periods. French ambivalence towards the US arises from similarity of conditions (economic and political strength) as well as from serious differences in political philosophy, cultural norms, and international policy, and from the constantly increasing power imbalance. Americans have the luxury of mocking and forgetting while France, like other countries, is unable to forget about the US because of the power imbalance but the same imbalance means that for the French the US is an interest "often verging on obsession" (Brenner and Parmentier 2002, 4).

A deeper historical foundation for Franco-American relations will be provided in chapter 3, but these points serve to illustrate the shaky foundations on which George W. Bush and Jacques Chirac built a fleeting friendship prior to the eruption of tension over US plans to invade Iraq. Both leaders were on the Right in their respective countries, a fact that perhaps explains why Chirac was the first foreign leader to congratulate Bush on his election in 2000 and the first foreign head of state to visit the site of the World Trade Center in 2001 (TF1 8:00 news, Nov. 3, 2004). France supported the US invasion of Afghanistan to overthrow the Taliban, as well. But the US plans to invade Iraq were met with great skepticism. French leaders contended that the invasion was unjustified because (a) Iraq had no weapons of mass destruction, (b) Saddam Hussein had no ties to al Qaida, (c) further arms inspections were needed, and (d) the planned invasion was in violation of international law. By 2005 the majority of Americans agreed with the French that the invasion was ill-advised (Roper 2005)[6] but antipathy towards the French remained. If anything it had solidified as a result of American discouragement with conditions in Iraq and a general sense of defensiveness. Some observers go so far as to characterize US-France relations as a "zero-sum game," implying that the two countries cannot reach mutual satisfaction on any issue (Davis 2003, 49). This is certainly an exaggeration, as the US and France share many of the same goals with regard to political stability, democracy, and economic development. It makes more sense to describe the current

6 To the question "Do you think the U.S. (United States) made the right decision or the wrong decision in using military force against Iraq?" 44 percent of respondents answered "Right Decision" while 50 percent answered "Wrong Decision" and 6 percent did not know or refused to answer (Roper 2005).

problem as a fundamental disagreement about the rules of the game, and specifically a split on the comparative utility of hard and soft power.

Ongoing Construction of the European Union

An intriguing consequence of the asymmetry in power between France and the US is that the more arbitrary and imperious the US appears the more French leaders benefit as the leading force behind the strengthening of the EU. Dominique de Villepin, France's Prime Minister and former Foreign Secretary and Interior Secretary, frames Europe's role in the following way: "We must now be aware that Europe has responsibilities. In a world marked by crisis, by difficulty, and by a profound disorder, Europe has a central role to play. No power, however great, including the United States, can assume sole charge of the world, carry the vision and construct a new global equilibrium. Europe must be the partner, the pole of stability and excellence that permits the world to move forward on two feet" (de Villepin 2003, 151). His argument owes a debt to Nye as he calls for a blend of two models of leadership – *le requin* (the shark) and *la mouette* (the seagull). The terms "crisis," "difficulty" and "profound disorder" (*la crise, les difficultés, un profond désordre*) evoking the world's need for France are meant to suggest something quite different from the Bush administration's "axis of evil." Instead of dividing the world into two camps, de Villepin evokes the image of disorder, which in turn suggests that the US's shark-like foreign policy has imposed chaos, and order can only be restored through a mix of the shark and the seagull. In a second metaphor – two feet – he envisions different roles for the US and Europe in precisely the way Kagan dismisses. He would consider Kant and Hobbes equally valid. His vision of Europe as the counterbalance to the US is not merely an expression of anti-Americanism. It is a call to Europeans to unite and exploit an opportunity created by the sclerotic hardening of American power. As we will see in the next chapter, de Villepin envisions a global role for the European conscience.

Although not yet united as a confederacy, let alone a federal government, Europe is arguably the most likely challenger to US supremacy over the next few decades – China, Japan, India, and Russia being handicapped in more significant ways (Nye 2002, 17–35). Europe's primary challenge at this point is its "size," shorthand for the heterogeneity created by the breadth of its current boundaries: "the prospects for a strong federal Europe may have disappeared when the original six countries agreed upon expansion that included Britain and parts of Scandinavia" (Nye 2002, 31). But protracted shark-like behavior on the part of the US – unilateral military action, aggressive trade policies, and general intransigence – could accelerate the coalescence of even this overly large Europe as a political power (*Europe-puissance*) rather than simply an economic space.

The motifs of Védrine and de Villepin demonstrate that French intellectuals and leaders do not simply object to US comportment; they *use* US comportment as a political resource supporting calls for a stronger Europe. The words of President

Jacques Chirac also demonstrate how national objectives and regional constructions of identity flow together: he argues that the goal is "to reconquer a sovereignty that in fact has already vanished" at the hands of US-led globalization (quoted in Brenner and Parmentier 2002, 31).

The concept of balance is crucial to the French and to their dream of a strong Europe in a peaceful world. The future they hope for (speaking broadly for the moment) is not just a global balance of power but a balance of cultural influences that would possibly prevent the ceaseless diffusion of American culture and the threat of cultural homogenization. In a February 11, 2005 interview on the Euro news Channel, French Prime Minister Jean-Pierre Raffarin placed the project of strengthening Europe in the context of "Anglo-American" cultural dominance of the world. "When I see the cultural diversity that exists today, I feel that we must defend it, and we need Europe, because otherwise we are going to live in a society with a single model, the Anglo-American model. We are in favour of diversity" (Raffarin 2005). Raffarin advocated the development of Europe as "An organisation that places greater emphasis on political responsibility, with a presidency, a Commission with real responsibilities that is accountable to the Parliament, a Parliament with real responsibilities, a true democracy with real political power" (Raffarin 2005).[7] So in the most troubling of paradoxes, Europe would counterbalance the US by becoming more like the US.

Without reference to the EU, France's geopolitical significance and particularly its relations with the US cannot be understood. The EU, arguably the strongest supra-national entity in the world, still lacks the executive centralization of a federal state. A shift towards confederation failed in 2005 due to a French referendum but this event must not be misread. The French "*non*" reflected displeasure with the economic liberalism of the current draft of the European constitution rather than a rejection of the EU-strengthening project, per se (Beaudin 2005, 43).

The French have an unparalleled legacy as architects and visionaries behind the European project, including such figures as Jean Monnet, Robert Schuman, and Valéry Giscard d'Estaing, yet bitter divisions on the direction of Europe remain in the French electorate. Under the leadership of the French Socialist Jacques Delors, President of the European Commission from 1985 to 1995, the EU evolved toward "embedded neo-liberalism," a combination of social welfare elements arising from local governments with free-market policies of the European Union. It is telling that Delors and his entourage sought to protect "the European model of society," meaning the role of government as a check on the power of capital and a source of far-reaching social protections. The French approach to development of the EU has therefore manifested socialist leanings and is at odds with German, Dutch and

---

7 This agenda was rejected by the French people in May 2005, and plans for a European "counterbalance" in place of the economic "space" are currently on hold. Progress depends on satisfying the Left that Europe will not be constructed on the free-market model of the UK or the US.

British visions of the EU. Internal power imbalances within Europe exacerbate these differences of political opinion by adding differences of interest.

> France sees in Europe an enlarged France with aspirations to counterbalance American power; Great Britain assigns Europe the role of America's privileged partner; Germany wavers between a pacifism inherited from the Nazi disaster and the desire to establish itself in a continental federation; the new member states see NATO as their guarantee of security more than Europe; Italy and Spain see in their relation with the United States an insurance against the always dreaded board of French and German directors, or even French, German and British.
>
> (Toulemon 2005, 94)

For the majority of the French, social protections are a European legacy deserving of protection and reinforcement even as Europe liberalizes its economies to meet the challenge of competition from the United States, China, and other regional groupings. The success of the EU in producing economic coordination is not matched by similar success in generating a regional social protection net, however, despite pressure from the French. The quest for economic competitiveness has driven the shift toward a laissez-faire approach. This shift was embodied in the European constitution and illuminated its defeat by French voters in spring of 2005.

Special interests of the French in the EU compound their unusually socialist vision of the project. For the French the European project has a special promise as indicated by Hilary Winchester:

> Much of the credit for the organizations of European unity must go to France; not only were individuals, such as Jean Monnet, influential in setting up the early European structure, but France as a whole adopted an expansive view of European unity, in which a strong Europe was viewed as an extension of French national interests.
>
> (Winchester 1993, 239)

France's desire to sacrifice its own sovereignty for the sake of a European confederation must be understood in light of the leading role that France has played in the current coalition and would almost certainly continue to play in the hoped-for confederation. In a political sense, the French would expect to see great returns on the investment of political capital in a strengthened Europe.

Comparison reveals rather intriguingly that with new members added in 2004, the EU-25 has a population 53% higher than the US, while its export revenues are 42% higher and its gross domestic product is virtually identical to that of the US (CIA 2006a; 2006b). Nonetheless, Europe's political and military power in the global system are insufficient to challenge the US and will remain so unless a stronger European political framework with a centralized executive branch can be created. As a market *space,* Europe is the US's match, but as a political *place* it lacks the coordination and unified vision necessary to rival the US military or check American political power. This remains the case although the EU member states individually are among the world's strongest states.

France's important position within the EU adds a further justification for studying French opinions about the US. Since 2004 the EU has been an open market linking 25 states (called the EU-25 when distinguished from its earlier boundaries) which is regulated by six shared governmental bodies based in Brussels. Although forged out of compromises between three European transnational classes (van Apeldoorn 2003) the dominant class has consisted of economic neo-liberals who support the lowering of trade barriers within Europe and also lowering trade barriers between Europe and the rest of the world. This class has made a few concessions to two other classes: a neo-mercantilist class advocating free trade within Europe but protectionism in the face of non-European markets, and a social democrat class attempting to build strong social protections into the framework of the EU, but "this incorporation is done in such a way that these concerns are, in the end, subordinated to the overriding objective of neo-liberal competitiveness" (van Apeldoorn 2003, 160). However, these international classes also map onto particular states when one views their predominant geopolitical motifs and actions, and French leadership in forging EU consensus does not indicate political consensus within France regarding how "social" the EU should be.

While the constitutional defeat of 2005 driven by factions on the Left and Right in France does not signal the end of the EU, the way forward is not clear. The French are positioned awkwardly vis-à-vis the European project, as both the primary drivers and the primary spoilers. For the present, the existing patchwork of treaties put in place over the past half century must continue to guide economic and political processes in the EU, in the absence of a constitution or real executive branch. Nonetheless, Europe perhaps offers the world a most elusive gift, a geopolitical motif:

> Confronted by two American visions of globalization, that of a pacified world at the "end of history" offered to us by Francis Fukuyama, and that of a world condemned to the "clash of civilization" with which Samuel Huntington menaces us, apparently opposite but each of them too narrow and dogmatic [*trop systématiques*], the Europeans, with this mix of skepticism endowed them by their overly rich and very old historical experience, and a recent but sincere attachment to international institutions, are well placed to promote a world order acceptable on every continent.
>
> (Toulemon 2005, 95)

To promote its more social vision of world order France seeks a more powerful regional framework, in (but not necessarily coextensive with) the EU.

Europe Puissance

The idea of a regional coalition with more far-reaching powers and influence in the global arena than those of the existing European Union (EU) is often couched in terms of a *Europe-puissance* (Europe-power). The realization of Europe's political and military weakness in the global arena, *as well as* its potential power measured in both cases against the US, has lent gravity to the discussion of *Europe-puissance*.

What is envisioned would be a new kind of political entity midway between a confederation of sovereign states and a federal republic (see Figure 3.2).

Since World War II European nations have voluntarily given up more sovereignty than other nations in the world. To strengthen the EU, European nations, France included, would have to give up even more of their sovereignty. They would, however, retain a level of sovereignty greater than that of the individual states that make up the United States. One reason for moving in this direction is to create a common defense. Another is to strengthen civil society and address a perceived "democratic deficit" by installing new forms of regional government or *governance*. The gap between the actual and envisioned autonomy of Europe raises questions about the utility of NATO, an alliance which most French observers consider to be obsolete and enchained to US interests. *Europe-puissance* implies strengthening executive and military unity to match the legislative and administrative coordination already afforded by the EU.

The Treaty of Nice, signed by EU member states in 2001, broaches the topic of cooperative defense in the most tentative possible manner: "The common foreign and security policy shall include all questions relating to the security of the Union, including the progressive framing of a common defence policy, which might lead to a common defence, should the European Council so decide. It shall in that case recommend to the Member States the adoption of such a decision in accordance with their respective constitutional requirements" (European Union 2001, 7). The treaty goes on to specifically acknowledge the pre-existing commitments to NATO of "certain Member States": "The policy of the Union in accordance with this Article shall not prejudice the specific character of the security and defence policy of certain Member States and shall respect the obligations of certain Member States, which see their common defence realised in the North Atlantic Treaty Organisation (NATO), under the North Atlantic Treaty and [shall] be compatible with the common security and defence policy established within that framework" (European Union 2001, 7).

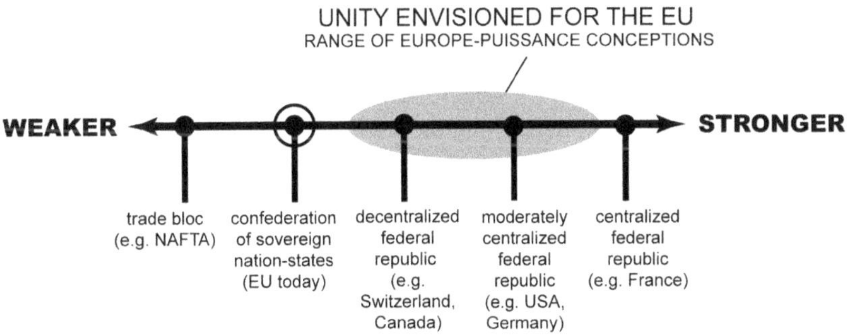

Figure 3.2 The envisioned unity of the EU relative to the existing unity of the EU and other existing entities. The EU envisioned by many European academics would take the form of a decentralized to moderately centralized federal republic. Author's diagram.

This language struck many "no" voters in France as making official the European dependency on NATO. Presumably for the French voters to accept an EU constitution it must explicitly distance EU military forces from NATO control.

In addition to executive and military unity, *Europe-puissance* would have to address the widespread perception of excessive distance between the European government and the European people (who still primarily perceive themselves as French, Germans, Italians, Walloons, Basques, and so on). This concern is often couched in terms of "democratic deficit," to the point that the term "has become a mere catch-phrase for everything that is democratically wrong with the Union (elitism, public apathy, secrecy in the Council of Ministers, etc.)" (Coultrap 1999, 107; also see Lord 2001, 642; Meadowcroft 2002, 181). Certainly, the wide range of complaints it encompasses makes the term "democratic deficit" somewhat vague, yet its prevalence clearly indicates that the EU is perceived by many Europeans as an inadequate framework for democratic decision making. The term indicates a failure of the existing structures of EU decision-making to be sufficiently transparent, responsive, powerful or motivating. In light of the unprecedented nature of the political construction called the European Union, this observation is perhaps not surprising. Most state democracies have suffered from "infantile health troubles" (Nowina-Konopka 2003, 2) and an institution barely 50 years old (even if one includes the European Coal and Steel Community, the European Economic Community and the European Community) would not be entirely past its own awkward phase. Even an author who defends the democratic robustness of the EU acknowledges the veracity of certain complaints about the deficiency of its democratic processes in comparison to the states that make it up (e.g. Moravcsik 2001). He points out, however, that no actual democracy matches philosophical ideals, and therefore democracies should be judged according to the "constraints imposed by the 'second-best' world of the specific case in question" (Moravcsik 2001, 347). To understand the French stance relative to the EU, however, it is necessary to consider the parameters of a "democratic deficit" debate in relation to globalization, US power, and idealized views of democratic process.

The problem with transparency amounts to the contention that those who make decisions in secret (such as the Council of the European Union) cannot be held accountable by the citizenry; decision makers should be forced to reveal and explain the reasons behind their decisions if the power delegated to them is to be legitimate (Majone 1998, 21; Lord 1998). Some dismiss the idea that greater transparency can solve the democratic deficit of the EU. Even a transparent process in the hands of a small set of representatives hides from people the true range of political possibilities and the ideological positions behind decisions (Magnette 2003, 150–57). The European public has no chance to review EU decisions until they seem complete – when the only acceptable responses are "yea" or "nay" and European governments actually work to separate European issues from partisan struggles at the national level (Decker 2002, 260). What is needed, on this account, is either a stronger dependence on elected EU representatives in more than just the European Parliament, which in any case is not vested with the power to control even the legislative process (Decker

2002, 261), or else direct participation by the European populace in debates over the issues affecting Europe. The latter would require much greater interest levels on the part of citizens (Decker 2002, 259; Majone 1998). In the current political climate, over half of eligible European citizens are not motivated to make even the small sacrifice of time and effort required to vote in European Parliament elections. This problem with motivation could potentially be addressed by politicizing the EU (Magnette 2003) and a step in this direction would be the emergence of pan-European political parties in place of the plethora of parties that are presently defined at the national level (Coultrap 1999, 115). Another step would be the generation of simpler and more dramatic EU politics centered on issues that move ordinary citizens, rather than the subtle and complicated issues that are pertinent only to elites. A third step would be the creation of an elected president similar to that of the United States (Decker 2002).

The public participation envisioned here fits within a larger agenda to promote *transnational* or *supranational* forms of governance. These terms stand in opposition to an *intergovernmental* form of decision making like that which currently dominates within Europe (Decker 2002; Grewal 2001; Rumford 2001; Schlesinger 1999). The distinction lies in whether the primary actors in Europe are the interacting governments of states, on the one hand, or groups and individuals who happen to reside in more than one state, on the other. In an intergovernmental system, groups are defined nationally (for example German environmentalists or French workers) and such groups address international problems by bringing pressure to bear on their respective states. The state may, in turn, change its domestic and/or foreign policies and place pressure on other states to enact domestic and/or foreign policies that are in the interest of broader groups. This hope is generally unfulfilled, however, as the priorities of intergovernmental relations are generally defined by nationalistic agendas. Transnational or supranational governance holds up the promise of transcending national and parochial concerns, creating a body of *united states*.

In the case of transnational or supranational decision making, citizen groups form and mobilize across state borders. Such political formations transcend the system of state borders and consequently have the power to shape the decisions of various states both individually and collectively (Taylor 1995). Transnational social formations employ transnational communications of a particular sort. Here we move towards a discursive conception of democracy the origins of which can be traced most directly to Jürgen Habermas, whose analyses of civil society and rational deliberation (1989; 1995) have provided the means for conceptualizing politics beyond the state. "If, in discussing their future, the generally privileged citizens of our region wish to take the viewpoints of other countries and continents into account, they will have to deepen the European Union along federative lines so as to create, as citizens of the world, the requisite conditions for a global domestic politics" (Habermas 1999, 47). Habermas' argument is that the procedural framework for just and moral decision making is something that states increasingly cannot provide on their own, due to the erosion of borders by the proliferating flows of capital. American-style neo-liberalism overcomes local calls for equality and security so a union of states is better equipped to confront the colonization of the lifeworld and recreate the basis of civil society.

> But if the nation-state is reaching the limits of its capacities in the changed context defined by global society and the global economy, then two things stand and fall with this form of social organization: the political domestication of a capitalism unleashed on a planetary scale, and the unique example of a broad democracy that works at least reasonably well. Can this form of the democratic self-transformation of modern societies be extended beyond national borders?
>
> (Habermas 1999, 47).

Habermas argues in the affirmative and most effectively presents the position of pro-EU social democrats. Of particular interest is Habermas' interest in "deepening" Europe which on his account has more to do with building responsiveness to citizens through the expansion of civil society to a transnational space than with building Europe's political or military stature in the world system.

This discursive, transnational approach to politics emphasizes the notion of *governance* – a realm of politics at once "below" and "above" formal, territorially-bounded practices of *government.* Governance is "the exercise of authority with or without the formal institutions of government," and it is *multilayered* insofar as it "includes important local, substate regional, suprastate regional, and transworld operations alongside and intertwined with national arrangements" (Rosamond 2000 and Scholte 2001; quoted in Rumford 2001, 208). Whereas government is territorial in its origins and limitations, governance transcends territories; it defines networks and spaces within and between states. One essential Habermasian premise – the idea of democratic discourse as non-coercive and consensus building – is rejected by many proponents of governance, however, because it presupposes ideal rather than actual participants. Habermasian citizens are able to completely overcome their biases, an achievement that is, at best, rare in the real world. Cognizant of this fact, Delanty (1998) proposes that "dissensus" rather than consensus be the foundation of political communication.

Out of this interest in governance emerges a major concern: even as the European arena has opened up for new forms of transnational governance and pan-European pluralism, the actually existing framework of the EU has thus far, and to an increasing degree, been dominated by neo-liberal principles of free-trade. Indeed the decline of state power resulting from international cooperation in Europe has left more openings for profit-making than for decision-making. "Consequently, in EU discourse society has become thought of less as a realm embodying solidarity and community and more of a partner in governance strategies for economic growth and competitiveness" (Rumford 2001, 208). Europe appears to be converging with the American model as Social-Democratic values at the state level have consequently lost ground to the imperatives of growth. In the terms of the existing debate, then, what is in play is how "social" (read socialist) a Europe, how "deep" (read coordinated) a Europe, and how "intergovernmental" (read decentralized) a Europe.

The conflicting views can also be mapped onto the division between Atlanticist and Europeanist countries (Stahl et al. 2004), and particularly onto the philosophical differences between the French and by other member societies. As the leader of the Europeanist contingent, France has charted a path towards a *social* (Social-Democratic) Europe. This direction was most evident and successful during Jacques Delors' ten-year Presidency of the European Commission and arose, in turn, from de Gaullian nationalism (Cohn-Bendit 2005). The social Europe model lost influence with the end of the Delors Commission (Rumford 2001, 208) but Atlanticism did not necessarily grow. The European response to US plans to invade Iraq in 2003 has been widely discussed within France as a failure of the EU's Common Foreign and Security Policy (CFSP) but the fact that Germany and France stood together against the US plans is worth noting as a sign of deepening. Historically, German politics have conflated cooperation with Europe and cooperation with the US under the single rubric of *Westpolitik*, but Germany's stand with France against the US therefore suggested the beginning of a new era of Europeanism long awaited by French leadership (Cohn-Bendit 2005, 82).

Should *Europe-puissance* as the French theorists envision become reality, Europe would constitute a major political and economic (and possibly military) force at the global scale, rivaling the US. Precisely how intergovernmental or centralized the new entity should be is a matter of debate in France, with the primary stances ranging from a greater Switzerland (*grand Suisse*) to a moderately centralized federal republic, a United States of Europe (farther right or left in the marked oval in figure 3.5).

The timing of the French call for *Europe-puissance* during the Bush presidency is not a coincidence; shifts in the perception of the US have led to a loss of confidence in US leadership at the global scale, concerns with unbridled power, a sense that the US confuses global interest with self-serving American interests, and a desire to construct a counterbalance to the *hyperpuissance*. In essence, it is the overt and sometimes arbitrary display of strength by the US combined with the fall of the Soviet Union which generates throughout Europe, to varying degrees, a decline in the will to remain dependent on the US. Such calls are based in the explicit rejection of the idea that global stability requires a single dominant power, and draw on the opposing conception of global stability as impossible without what Jacques Lanxade calls a "balanced strategic partnership" (*partenariat stratégique équilibré*) (2004, 163).

Conclusion

Because of its strength, perspective, geographical position, historical role and global influence, France is a particularly fruitful source of geopolitical discourses to place under the microscope and look for clues to the shape of the 21st century. France has broken with the US more than any other ally in expressing opposition to US foreign policies. Most of the post-World War II period has been marked by the refusal to cooperate fully with Washington's plans and with the assertion of a French agenda for the future of Europe and international relations. In the context of the United Kingdom-France-Germany triad, France is least likely to see eye-to-eye with the United States while the other two are more easily swayed by the US and this situation is directly responsible for the attempts by the French to strengthen the ESDP so that European forces can operate independently from NATO. This endeavor is at odds with the dominant American conception of European forces as either fighting alongside American forces in engagements identified, planned, and commanded by the US, or undertaking post-conflict situations "clean up" after US-led battles to restore order, democracy and capitalism. The fervent desire of some Europeanists in France to detach from NATO obligations and the inertia of the half-century old defensive alliance create major ambivalence in France that has contributed to the failure to ratify the EU constitution. While American observers, particularly neoconservatives, express disapproval for the relative weakness of European militaries, an ESDP separate from NATO is not what they have in mind as the alternative to the status quo. France's leadership in attempting to create this new military pole will be fruitful precisely to the degree that American power appears arbitrary, misguided and self-serving.

If US policies are supported by France, as was the case with the 2002 invasion of Afghanistan, then France can help rally global support to the side of the US. Conversely, if France takes an opposing position its substantial economic and cultural resources enable it to exert an unusual influence over European perceptions and representations and to play a prominent role in the UN, potentially undermining support for the US. The geopolitical motifs employed by the French in representing the US therefore have far-reaching importance.

As two "leading" nations, the US and France promote rather different objectives for other nations to follow, and they envision international leadership in different terms. These differences lead to tension as does the rivalry that is inevitable between two nations self-defined by the same idea. "When one has designated oneself as the model and the emissary for a universal message of liberty and equality it is hard to tolerate another who takes on the same occupation" (Boissonnat 2003, 588). It is precisely what the US and France share, a certain idea of the nation as a model and emissary of freedom, that infuses the US-France relationship with tension. But the differences in historical geography and social memory orient the two national missions in different directions: "Bush reasons no differently than Napoleon when he justified his crusading impulses in the Middle East by the necessity of offering democracy, religious freedom and the market to new people" (Boissonnat 2003, 588). Far from excusing Bush, this comment reveals a problematic historical lag since the lessons of colonialism were hard won: "the messianic version of the nation and its role in the world… connects American nationalism of today to nationalisms of the past that were also 'dissatisfied,' those of Germany, of Italy and of Russia, rather than to the satisfied patriotism of the status quo, characteristic of the British…" (Lieven 2004, 23). Nationalist attitudes have developed into totalitarianism more than once in Europe's history and that haunting memory suggests that the US will fail in its "dissatisfied" version of the civilizing mission, dragging down its closest allies. So it is best not to cleave too close.

Because of its policies as well as its aggressive rhetoric (ranging from bumper stickers to political theory to cabinet-level announcements) the US now is perceived as an empire, although without the stewardship ethos that characterized European empires in the colonial era (Joxe 2004a; Conesa 2004). Critiques of the American Empire are now launched across the Atlantic from an emerging group of united states that seem handicapped by divergent interests and an undeveloped frontier to the east. The geography intriguingly mirrors that of the colonial United States as does its potential to emerge as a world power if it can overcome internal fragmentation. Favorable views of the US among European citizens have fallen by fifty percent during the Bush administration (Pew 2003) and European intellectuals are already calling for a concerted effort over the next decade to prevent the world from becoming increasingly dangerous, intolerant, ugly, rude and polluted as a result of US manipulation. The most important resource of these newly united states may be their soft power.

In short, after two centuries the tables are turned on the United States of America. But the parallel only carries our analysis so far. Europeans would be less concerned about the United States, and vice versa, were the two continents as divided by the Atlantic as in the colonial era. But the adoption of countless technologies over more than 200 years has brought all human groups closer together in space-time, with noticeable political impacts. It is to the media that we must therefore turn at this point.

Chapter 4

Scholarly Debate: The Emerging Motif of Counterbalance

> We should therefore give up looking at all democratic peoples through American spectacles and try at last to see them as they actually are.
>
> de Tocqueville, *Democracy in America*, p.456

What constitutes an "event" varies from one communication medium to another. Scholarly research as a medium of communication includes a deep sense of history and the ability to survey a multitude of events rather than simply the most recent events. A scholarly publication has certain affordances including a historical long view in which *political culture* is of greater interest than fleeting events like elections (unless history should prove that the election to have signaled some momentous kind of change). So from the point of view of French political scientists, Bush administration policies indicated trends in American foreign policy that were evident long before the Bush administration; the 2004 election was scarcely an event and Bush's re-election merely pointed to four more years of the same.

A few articles appeared after the election in scholarly periodicals (Julliard 2005; Michelot 2005; Froment-Meurice 2005) but French scholars generally inserted the election into familiar arguments that had been made over the course of many years. To make up for this absence, political analyses by American observers were reprinted in translation. Within other media such as newspapers, as we shall see, the election *was* an event, for reasons that have everything to do with the affordances of a different medium, but the affordances of scholarly journals suggested that any single election could not, in itself, be particularly important or informative so the re-election was essentially a non-event from the perspective of this medium.

Therefore the contribution of scholarly debate to the symbolic construction of the 2004 American election must be situated in a longer time frame. In this chapter we consequently survey French scholarship over several centuries to discern the way in which representations of the US are bound up in representations of Europe and vice versa. The US provides French intellectuals with a vehicle for envisioning France's future and an incentive for the "European project" – the motivation and the model for a united Europe. In doing so the US is seen as having great virtues and great flaws. In drawing on these themes, intellectuals would help shape representations of the 2004 election in other French media such as newspapers, television and the Internet.

Behind the communications in this chapter are those authors whom Ó Tuathail and Agnew (1998) call "intellectuals of statecraft." The geopolitical writings of these pundits, political scientists, and politicians are goods crafted for an audience segmented by education, class, regional perspective, and political leaning. Such authors are able to maintain their role and livelihood, their *author*ity, because of people's interest in consuming representations that confirm their pre-existing worldviews. Despite their authority, intellectuals of statecraft are quite often read by only small circles of intellectuals who have little opportunity to directly affect policy or shape public opinion outside of the act of encouraging and subsidizing the flow of symbols they find congenial by purchasing books and journal subscriptions. Yet at certain times these authors can shape public policy, affect legislation, mobilize popular movements and capture large amounts of media attention. Usually this happens when certain events destabilize worldviews and require some effort to "claw back" the events into familiar media frames (Fiske and Hartley 1978, 87). J.L. Richardson describes key events in the history of a nation as "moments of high drama which can illuminate the political landscape ... like a flash of lightning" (Richardson 1992, quoted in Dijkink 1998, 198). But of course the drama itself provides little illumination; rather it is the motifs provided by scholarly debate that pick out the high points and the low points, envisioning transformations and continuities in the political landscape, re-charting a terrain reworked by tectonic activity.

The motifs related to a particular event have a history marked by borrowing, translation, rejection, and resurrection. To understand the French scholarly representations of George W. Bush in 2004 we must therefore reach back in time, not just to the beginning of that administration, but to the foundations of Franco-American relations as illuminated in the writings of French scientists, historians and politicians. Subsequently we return to the present moment and survey representations of Europe and the United States by contemporary intellectuals of statecraft, concentrating on themes of power and scale applied to Europe, the essence or character of the US, and relations between the two.

Anti-Americanism?

The French have been fascinated for more than two centuries with the savage and untamed character of the American landscape and society. The brutality and rough edges evident in everything American from teen music idols to the death penalty contrast with the French orderliness and sense of social obligation. From the vantage point of a social order rooted in a Catholic heritage, a political culture of centralized management, and centuries of relatively unchanged rural life, America presents a contrasting image of disorder made oddly more threatening by its paradoxical legacy of having restored order to a war-torn Europe.

According to Laurent Cohen-Tanugi (2003, 56), "critique of America's foreign policy has constituted ... an identifying trait of the European intelligentsia since at least 1945, France being the champion in all aspects of this endeavor." John Keiger writes that "If ever there was a mythical beast that stalked the minds of the French in the twentieth century it was the 'Anglo-Saxons'," and he argues that France has defined itself during the 20th century as much by what it is not as by what it is, its antipode being "all things Anglo-Saxon from the English language to 'la mal bouffe' and the 'MacDonaldization' of the world" (2001, 160, 183). While these pronouncements are a bit glib, other scholars have indeed identified and documented a persistent strand of critique directed at the US from French intellectuals. Jean-François Revel's *L'obsession anti-américain* (The Anti-American Obsession) notes that: "the primary function of anti-Americanism has been, and still is, to tarnish the image of [economic] liberalism in its supreme incarnation" (Revel 2002, 31). This critique emanates from French intellectuals on the Left, such as Jean-Paul Sartre, as well as those on the Right, such as André Malraux (Revel 2002; Keiger 2001, 181).

Most notable among the French scholars of anti-Americanism is Philippe Roger (2002) whose *L'Enemie américain* exhaustively documents anti-Americanism in French writings from the time of the English colonies. His diagnoses of anti-Americanism are a bit overdrawn in some cases, since the subjects in question range from the poisonous flora and fauna of the New World, to the manners of New World colonists, to American military and economic policy. Furthermore, some of the attitudes Roger catalogs as "anti-Americanism" could more accurately be described as anti-modernism since America has been employed by French intellectuals as a symbol of modernism. Of greatest concern is that Roger includes as "anti-Americanism" many expressions of concern about military aggression and economic competition from the US. Surely the anxiety of the French in this case has as much to do with a concern for geopolitical balance of powers as with antipathy towards America, per se, since any other highly assertive and powerful country might well inspire the same kinds of anxiety. Nonetheless, Roger admirably documents French views of America, and the following account owes much to his findings.

It is clear that for reasons both intrinsic and extrinsic to America itself French observers have been alternately concerned, alarmed and offended by the view across the Atlantic. In Roger's words:

> America often exerts a strange power over the [European] imagination. Anything seems possible on a continent where the line between the believable and the unbelievable sometimes dissolves. The distance of this continent appears to disturb the judgment and introduce an emotional element that often alters the level-headedness of observers. The image of America therefore has always been formed of hope and anguish, political utopias and philosophical visions that are often contradictory.
>
> (Roger 1996, 21)

These responses, while too ambivalent to be called "anti-Americanism," are nonetheless the foundations for the current French representations of the United States.

Early transatlantic observations

Eighteenth century observers such as Georges-Louis Leclerc, Comte de Buffon, commonly known as Buffon, and the Dutchman Cornelius De Pauw tended to view the North American air and soil as unhealthy, a hypothesis seemingly supported by evidence of the stunted and sluggish condition of American plants, animals and people. From second-hand accounts and ad-hoc theories this appeared to be an unordered place, a wild and chaotic landscape rife with poisonous plants and venomous animals, where species could not develop to their natural size or vitality. Although Buffon was a brilliant mathematician and biologist, his ecological theory linking the environment of the New World to dwarfism and lethargy, both human and animal, was clearly based on misinformation and prejudice (Roger 2002, 21–57).

Opposed to this was the noble savage myth, which originated in 16th c. France with Ronsard and Montaigne and reached its apogee in the 18th c. with the artistic and literary visions of Rameau and Rousseau. The noble savage was a "simple and good" version of humanity "free of the prejudices of European society, leading a happy life in a nature that was consistently benevolent" (Roger 1996, 21–2). For a time the Native Americans were invested with these qualities, then as they were driven into the west, the white settlers who drove them out were invested with similar virtues. French authors who extolled the apparently virtuous and frugal society of Americans in the noble savage tradition included Abbey Raynal, Turgot, Condorcet, and Voltaire.

The next generation of French observers hailed the American Revolution and drew inspiration from it as they played a role in their own revolution, although some saw the US as a disturbing example and others, such as Simon Linguet, were concerned that an independent North America would become strong enough to take over Europe. But disdain soon dominated rather than such fear since French visitors found the new Americans sorely lacking in manners. In a sense they were right. These Americans did lack *French* manners, and the majority of them valued manners in general less than other measures of personal worth such as material prosperity, practicality, piety and diligence. Still, their disdain did not prevent many French republicans from upholding the American political system as a model for French democracy.

The French Revolution intervened in these perceptions in two ways. First, it reduced the need for foreign political models. Why look abroad for political models once the French had overthrown their monarchy and created a democracy of their own? Second, Americans demonstrated a maddening degree of indifference to the onslaught from the various monarchies of Europe from the time of the Revolution through the Napoleonic era. Indignation fostered contempt, as the absence of (French) manners and the refusal to repay its military debt to France seemed to prove allegations that America was a base, greedy, and self-centered society.

The characteristic American preoccupation with business and profits further reinforced this image. For the French, civilization involved the flowering of arts, literature, philosophy and cuisine, while it appeared that such things scarcely existed in the US. The French diplomat, Talleyrand, living as a dissatisfied expatriate in Philadelphia during the last decade of the 18th century, gave particularly scathing accounts of American life. His reduction of the US to "thirty-two religions and a single meal" indicated discomfort with American sectarianism as well as an offended palate (Roger 2002, 70). The supercilious tone was firmly established by the turn of the 19th century. Volney expressed the idea reminiscent of Buffon, that the US had regressed politically; Bayard and La Rochefoucauld-Liancourt expressed disappointment with the Americans' conversational skills, their lack of interest in intellectual speculation, and their preoccupation with money; Stendhal mourned the lack of opera; and Baudelaire added: "what an odor of shops" (Roger 2002, 74, 61). Aside from fine cuisine, fancy clothing, clever conversation, and classical music, the French ideal of Culture is a common good, a shared legacy for all citizens, a purpose to secular social life beyond self-serving and materialistic pursuits. American utilitarianism and practicality offended the champions of Culture on the Right while the country's competitive ethos disturbed the egalitarians on the Left.

In the first third of the 19th century, the French admirers of America for a time outnumbered detractors, although the tenacious strains of critique continued. America still presented the image of wildness, space, the savage and the exotic since Indians were still present in the landscape along with pioneers, mingling the rural idyll and the noble savage with a certain air of grandeur and mystery. On the political side, the US and France were both democratic republics and the US presented an image of a restricted government and a market-oriented society that appealed to the "liberal" French observers on the right wing of the political spectrum (the French term "*libéral*" being opposite in its connotations from the English word "liberal"). But by the 1830s, French opinion turned critical again as visitors to the US became more numerous and their impressions confirmed the negative impressions of earlier observers. Once again it appeared that Americans placed material gain above other human and social values and for the first time (surprisingly) slavery was widely criticized. Whether as slaveholders, bookkeepers, or eccentric millionaires, Americans seemed to be driven above all by avarice. They likewise seemed oblivious to attractions in life other than fighting – as the violent nature of American society was reviled at this time, as well. Here Alexis de Tocqueville intervened in the discussion with his remarkable *Democracy in America*, adding no small amount of support to both the critique and the celebration of America.

De Tocqueville and the crystallization of transatlantic critique

The writings of Alexis de Tocqueville have been treated in the US as the quintessential 19th century European perspective on the US, a view which overlooks Tocqueville's unusually positive tone compared with other European observers of the period. By the late 19th century many French writers excoriated his "sugar coated" portrayal of the United States (Roger 2002, 90). The scholarly nuances in this brilliant work were easily manipulated by writers who wanted only to hear his negative views; the champion of American democracy seemed most authoritative when his words reflected badly on the US. What Tocqueville did in his two volumes, however, was not simply document the character of American political and social life but persistently and persuasively link one to the other. Democratic governance had shaped the American people through their institutions and daily experiences, he argued, to the degree that American attitudes toward religion and science, manners and money, arts and literature, all took peculiar forms essential to democracy and therefore largely unfamiliar to Europeans. America should be of interest to Europeans, he suggested, not simply as a curiosity but because Europe itself was democratizing and the US indicated the path that lay ahead, with all its promises and its pitfalls. Tocqueville did not deny that American society had faults and weaknesses, but he was willing in most cases to accept these as the inevitable costs of democracy.

Tocqueville is one of the great early proponents of democracy, a learned and perceptive observer, and a brilliant analyst who interpreted America's cultural peculiarities as elements of an overarching cultural complex in which each part is causally related to the others. Tocqueville's belief in the systemic interdependence of various elements of society certainly foreshadows 20th century political science even when his interpretations clearly display a 19th century sensibility. His motley and compromised American utopia marks a kind of upper limit beyond which few pro-American observers in France are willing to go in their praise. His critiques of American society therefore deserve attention not only for their own sake but also for their role in crystallizing the motifs by which America and Americans are represented in France.

These critiques touch on four general areas – economic, political, cultural, and spiritual – which we will consider in turn. On the economic plane he feared that oligarchy would come to occupy the same position of social dominance in democracies that was held by landed elites prior to democracy: "if ever again permanent inequality of conditions and aristocracy make their way into the world, it will have been by that door [leaders of industry] that they entered" (Tocqueville 1969, 558). On the political plane, his concerns turned on the question of political centralization, which he believed might increase ironically as a result of excessive egalitarianism (Tocqueville 1969, 673). On the third plane, that of culture, he offered rather dire pronouncements that have become grist for the mill of French anti-Americanism.

> The majority in the United States takes over the business of supplying the individual with a quantity of ready-made opinions and so relieves him of the necessity of forming his own. So there are many theories of philosophy, morality, and politics which everyone adopts unexamined on the faith of public opinion … If democratic peoples substituted the absolute power of a majority for all the various powers that used excessively to impede or hold back the upsurge of individual thought, the evil itself would only have changed its form. Men would by no means have found the way to live in independence; they would only have succeeded in the difficult task of giving slavery a new face. There is matter for deep reflection there.
>
> (Tocqueville 1969, 436)

This is tyranny of the majority and the French see it far more easily in America than in their own society. Other cultural concerns included America's lack of significant scientific, literary, or artistic endeavors, the corruption of the English language, and the greed, hedonism, shallowness, and political apathy of Americans (Tocqueville 1969, 454–6, 480–82, 615, 671).

Rounding out the critical aspect of Tocqueville's writings was a concern with the moral implications of democratic life. He noted somberly that: "One must not blind oneself to the fact that democratic institutions most successfully develop sentiments of envy in the human heart," and in a similar vein, he concludes that equality, "while it brings great benefits to mankind, opens the door … to very dangerous instincts. It tends to isolate men from each other so that each thinks only of himself" (Tocqueville 1969, 444). This moral concern converged with his fear of an American oligarchy, his concern with conformity, and his acknowledgment of crudeness (albeit understandable) in American manners to reinforce a familiar stereotype. The Tocquevillian critique cuts across the planes of economics, politics, culture and morality and serves as a key to interpreting the writings of subsequent French observers.

Political struggles and the distant mirror

As French democratic forces struggled against both local and foreign aristocracies during the 19th century, the US was upheld as a positive political example then subsequently critiqued depending on shifts in the winds of political change. Even in the more friendly phases, the idea of borrowing ran up against engrained prejudice: Americans were seen as members of the "Anglo-Saxon race" and therefore subject to the flaws and limitations previously identified in the British (Roger 2002, 238–69), and what is more, the troublesome Anglo-Saxon blood was now being mixed with the inferior blood of slaves, Indians and immigrants, which was sure to bring about social instability and cultural degeneration (Roger 2002, 270–89). By this point the US was no longer a kind of living history museum but the old themes of savagery, noble and otherwise, were recycled to give meaning to American immigration and urbanization. America's great cities, and above all New York and Chicago, manifested a kind of chaotic abundance and untamed quality, a profusion of new machines and diverse peoples that was unsettling. It seemed that the development of Culture and the transmission of civilization could scarcely endure in such a chaotic environment.

Meanwhile, and in a strikingly divergent way, 19th century visionaries from Napoleon Bonaparte to Victor Hugo took inspiration from the US as a model of the future of Europe:

> France has something wonderful about it as it is destined to die, but to die like the gods, by transfiguration. In the 20th century you will have an extraordinary nation. This nation will be great, which will not prevent it from being free. It will have Paris for its capital and will not call itself France but rather Europe.
>
> (Hugo, quoted in Boissonat 2003, 587)

This is the burden of the *mission civilisatrice*, the French civilizing mission that is global in scale and rooted in the enlightenment even if it was voiced by rather unenlightened leaders such as Robespierre, who declared in 1790 that "It is in the interest of other nations to protect the French nation because it is from France that liberty and well-being of the world must come" (quoted in Boissonat 2003, 587). At the beginning of *Fragile Glory*, an exploration of modern France, Richard Bernstein captures this French desire "to represent something beyond themselves, to light up the world, to glow with the torch of civilization itself" (Bernstein 1990, 3). In the minds and discourses of French leaders and cultural elites the tension between America-as-model and America-as-threat remains vital to this day. If the American social model is based on the practicality of the yeoman farmer and the calculating spirit of the urban businessman, the French model is based on a Catholic (or post-Catholic) sense of societal harmony, inclusion, and transcendence. Yet both countries would have their model be the world model:

> France, since the 18th century, would make of itself the bearer of a vast ideal, a bit like the United States. It has always presented its political project not as a pragmatic construct that permits people to live together while adapting to change, but as a pathway to universality, a stone for building a universal republic, a marvelous machine to bend egotism under a single law.
>
> (Colombani 2000, p. 17–18)

So America provided a model for Europe even as France was envisioned in Paris as the model of the world.

The legacy of the 20th century

French intellectuals of statecraft drew on the ideology of *mission civilisatrice* to justify colonial projects in Asia, Oceania, Africa and Central America well into the 20th century, even as the rise of the US seriously threatened this cherished ideology. The intervention of the US on behalf of France in both world wars produced less gratitude than the Americans are wont to believe, because of the details surrounding both interventions which pointed to self-interest rather than the US's vaunted principles as the underlying motivation. In regard to the First World War, the French were of the impression that the American expense in dollars was matched by the French payment in blood, whereas the US expected France to repay its war debt. France's response was to launch a European unification project between the wars, a project that scarcely paused during World War II. As early as the interwar years, anticipatory efforts toward European unification could be noted, although the European Coal and Steel Community would not be put in place until 1951. By 1920, Albert Demangeon, a geographer, was calling for European unification and explicitly casting the European project as a defensive response to the burgeoning power on the far side of the Atlantic:

> It appears clear that, on different territories and under various titles, the heirs of Europe are the United States and Japan. For a long time the Monroe Doctrine has marked the limits of European political ambitions on the American continent; the prodigious development of the United States in industrial productivity likewise imposes limits on the economic expansion of Europe; Latin America, long our commercial territory, gives in bit by bit to the Yankee attraction; even more, by a curious inversion of currents of influence, Old Europe opens up to young America as a terrain of colonization.
>
> (Demangeon 1920)

If the US was undermining the sovereignty of France it remained a model. Demangeon provocatively employed the phrase "United States of Europe" as early as 1928, in an article in *Annales de géographie* and the same term was employed a year later by France's Foreign Minister, Aristide Briand, in a speech to the League of Nations (Parker 1987, 148). Briand developed this idea into the Briand Plan for the European Union. Meanwhile another French geographer, Jacques Ancel, advanced the notion of *groupement* as a multicultural response to the German idea of *Raum*, implying an open cosmopolitan association of peoples rather than a closed, ethnocentric territory for a single ethnic group. French geographers of the early 20th century did not see France changing to become more European but rather saw Europe evolving to become more French.

After the Second World War, Europeanism gained currency in France, fed of course by Gaullist antipathy towards the US but also by popular sentiment. In a survey conducted in 1945, only 20% of the French population gave the Americans the most credit for the defeat of Germany in the war, while close to three times as many people (57%) gave the Soviet Union the most credit (with another 12% ranking Great Britain top among defenders). Explanations for this eastern orientation are complex: a persistent belief in the superior strength of the Soviet military, a political preference for socialism, and tensions between French citizens and American troops during the period of American liberation probably all came into play (Roger 1996, 45–7). And once again the post-war economic settlement caused friction (Roger 1996, 188). Certainly the late entry of the US into the war, the marginalization of de Gaulle and *France Libre* as part of the US's "Vichy gamble," the exclusion of France from important negotiations and particularly the Yalta Conference, and the management of postwar finances all contributed to the French lack of enthusiasm for the country that saw itself as the liberator of Europe (Langer 1947; Keiger 2001, 176–9).

Thus, when French statesmen Jean Monnet and Robert Schuman developed the European Coal and Steel Community after World War II, planting the seed that would grow into the European Economic Community and ultimately into the European Union, their efforts brought to fruition the idea of a regional coalition that had been developed by early 20th century French geographers anxiously looking across the Atlantic and copying America in order to be able to expand French civilization and resist Americanization (Parker 1987, 149; van der Wusten and Dijkink 2002). If Germany provoked anxiety, as well, this was a fear "that dare not speak its name" (Keiger 2001, 216). Representing the German threat overtly might increase it whereas denying this threat and reverting "to the old ploy of drawing Germany closer to restrict its freedom of manoeuvre" might neutralize it (Keiger 2001, 217). If the British seemed fully ensconced in the American empire, they might yet be inclined to consider uniting their government with that of France – an idea quite seriously, if briefly, considered in 1940 (Keiger 2001, 175).

Communists and others on the left still echoed long-standing critiques of American materialist values while those on the right were not yet free from the aesthetics of the aristocracy, including a visceral disgust at popular icons like Mickey Mouse, hamburgers, and rock and roll music. As US dominance increased on economic, political, military and cultural fronts, cultural globalization and homogenization became linked to Americanization in the eyes of European observers. On the cultural plane, America took on particular role in French discourses relative to matters of the "spirit," a word that is different but no less ambiguous in French than in English. As the Catholic Church lost adherents and French society secularized the source of the problem appeared to be Americanization. Roger (2002, 533) summarizes the French interpretation in the intriguing formula: "the Americans are falsely religious citizens in a falsely secular state." This phrase indicates a French perplexity with a people who paraded their religiosity while apparently placing material values topmost: America's "separation of church and state" was belied by the use of the Bible in courts of law and prayers in legislative sessions. However, the Protestant faith was variously described by French observers as gloomy, prosaic, lacking in mystery, dry, empty, cold, proud, and hypocritical – a religion hardly worth the name (Roger 2002, 532–7).

Finally, the US was in something of a "damned if you do, damned if you don't" situation, cursed by the French for foreign meddling and also for succumbing to isolationist tendencies (Revel 2002, 21–2). These foreign policy critiques were not necessarily incoherent. The increasing globalization of the 20th century had produced "a world where the [US] could no longer isolate itself," which essentially gave "new clothes to an isolationism which strove … less to avoid contacts with Others than to ensure that these contacts made no alteration in the character of the experience that it was the US mission to protect" (Melandri 2003, 28). The US was cursed both for intervening and for failing to intervene because of how and when it applied these contrasting approaches. Both approaches were applied in ways that demonstrated a certain attitude of supremacy, and without this attitude each could have been engaged differently, at different times and in different places. In fact there were many alternative foreign policy choices that the French would have chosen (or believed they would have chosen) if they were in the US's position.

Tensions between Washington and Paris subsided during the late 1970s and 1980s although France still strove for autonomy within the polarized geopolitical landscape of the Cold War era (Moïsi 1995, 65–6; Keiger 2001, 182). But the fall of the Soviet Union in 1989 was followed by a wave of aggressive trade liberalization emanating from Washington that French leaders such as Mitterrand and Chirac did not always meet with enthusiasm. This time the socialist tradition of labor activism and social democracy in France led to suspicions regarding US-backed free trade policies. The increasing power of the US to direct world trade and maintain its advantage was simultaneous with the decline of France as a world power. By the 1970s Valéry Giscard d'Estaing pronounced France a "middle-sized power" (Bernstein 1990, 12), initiating media reflections such as a 1987 cover story in *Le Point* with the title "Inquest into the Decline of France" (Bernstein 1990, 12). American power and French weakness appeared inextricably linked.

These negative judgments did not indicate a complete distaste for America. Indeed as a tourist destination and as a source for the borrowing of selected cultural elements, particularly fashions and music, America was more than ever an object of fascination. The early 1980s brought a phase of "Americanomania" (Keiger 2001, 183). Aside from its intriguing culture, the US was still a place of wild nature with wide open spaces and magnificent natural features such as Grand Canyon and the Great Plains. But a social dimension of wildness seemed to come with the landscape; this was also a place of misbehaving children, gun violence, capital punishment, rampant divorce, cut-throat economic competition, degrading poverty, and the all-powerful hand of the market. For both religious and secular observers in France such signs indicated a crisis of the "spirit" in America. Wildness of a social variety was concentrated in American cities, making them seem unnatural and monstrous. Jean-Paul Sartre compared Los Angeles to a "fat earthworm that one could cut into twenty pieces without killing it" (Roger 1996, 176).

The image of Americans as rude, crass, shallow and mercenary Anglo-Saxons remained present in French discourses while French observers saw the countries drifting even further apart in regard to their views on many things, including "the relative roles of individuals and society, the place of religion, the relationship between policy and ideology, the conception of law, and confidence in progress and science" (Andréani 2001, 296). It was a matter of differing philosophies: "There remains a bit of Rousseau among us, and not among them; we have different ways of conceiving of punishment – the death penalty, for example; differences in the way of understanding social security, etc." (Calvez 2004, 28). But it was also a matter of different aesthetics. An image self-consciously mobilized by presidential candidates targeting American populist sentiments, such as Reagan, Carter, and Bush, is that of the rough-hewn man of the American heartland – a plain speaking, down-to-earth, practical fellow who knows how to save a dime and who volunteers for the posse when an outlaw is on the loose, and above all (here even Clinton could compete) a guy who likes simple foods like hamburgers and jelly-beans. This appeal is as poorly received in France as it was well-received in the US, for reasons that should by now be clear. Thus American Presidents activate a whole host of associations deeply rooted in the relations between France and the United States. Their speeches, actions, and personal iconography are tailored in a way that produces discomforting associations, including primordial critiques of American manners, culture, and civilization.

Contemporary Critique

Most recently, from the beginning of the George W. Bush administration, a lack of diplomacy has undermined the US's soft power in the French political sphere. François Duchêne, a Europeanist of eminent credentials, wrote: "President Bush in effect gave to his rejection of the Kyoto protocol – which the Senate would not have ratified in any case – the most provocative tone possible" (Duchêne 2001, 603). The situation rapidly worsened after the brief "honeymoon" presented to the US by the 9–11 attacks (Colombani and Wells 2004). Donald Rumsfeld's caustic reference to "old Europe," and Condoleeza Rice's call to punish France bred ill will towards the US while strengthening France's soft power in Europe.

Fleshing out the contemporary period in French representations of the US requires a survey of recent intellectual writings. The sources for the following analysis are books by prominent French authors and articles published in recent years in scholarly French periodicals. The books I have chosen embody the thought of the most prominent and prolific authors currently publishing for the general French readership. I have avoided works that are directed toward an audience of specialists and are therefore limited in their influence. The periodicals I have chosen – *le Débat, Commentaire, Études,* and *Futuribles* – participate in French social, cultural and public policy debates. Collectively these journals published over fifty articles in the 2001–2004 interval relating the US-Europe relationship or the future of either the US or Europe from a political perspective. I excluded articles that were focused toward narrowly specialized audiences in areas such as economics, policy, or administration.

These periodicals provide a representative cross-section of geopolitical debate. *Le Débat* is a 25-year-old periodical covering history, politics and society, released six times a year by France's third largest publishing house, Éditions Gallimard (Gallimard 2005). It aspires to be "an ambitious review – one made to last" while avoiding a close association with a particular school of thought or political program such as typifies other French scholarly journals (Nora 2000). *Commentaire* is an eclectic review that appears three times a year and was inaugurated in 1978 with the purpose of "following the movement of ideas that, in France and beyond, condition the destiny of our societies over the long term"; it addresses political and social sciences, history, international problems "without forgetting art, literature and philosophy" (Aron 2005). *Études* is a Jesuit monthly founded in 1856 addressing news, current events, politics and society: "fundamental themes [are] treated with openness and seriousness. It is a pleasure to understand beyond the event" (Assas Editions 2005). Finally, presenting itself as "the primary futurology journal in the French language," *Futuribles* strives to provide readers with "a better understanding of the contemporary world, analysis of possible futures and associated risks, and the policies and strategies that may be adopted" (Futuribles 2002–2004).

These four sources were selected for their inclusion of significant in-depth analyses of France, Europe, and the United States and for the quality of thought demonstrated in their articles, as well as for their availability at the French National Library (BNF François-Mitterand).

From a space to a power

French intellectuals of statecraft currently manifest a range of constructions of France, Europe, the US, and the world, but most center on the single, overriding idea that within the space Europeans have carved out for free trade there must develop a stronger union. This new Europe would be a unified world power with the ability to respond to the capricious, reckless, and fundamentally self-serving actions of the US. The push for European unification reflects a set of interrelated objectives that can be found in many regionalist movements today: to regain control of national security in the post-Cold War era, to facilitate international cooperation, to respond to sectionalist movements, to avoid marginalization in the world system, and to better accommodate minorities and stateless peoples (Dodds 2000, 23).

French scholars in international relations and foreign policy are generally in agreement on the need for Europe to become something more than just *une espace* (a space) for free trade – to move toward being *une puissance* (a power) – although the geographical boundaries and social organization of this power vary greatly in conception. While it is clear that geographical scale and size of population benefit a trade zone and a currency, economic and demographic scale alone are not sufficient to create political strength (Bollaert 2001, 19). *Europe-puissance* is a necessary response to *l'hyperpuissance*.

> Paris, November 2, 2004. Without a doubt the American elections have never been so much in the media, the opinion polls so uncertain, the whole world hanging on the outcome of an election which, unless contested (they say 20,000 lawyers have been recruited by the candidates) will determine the future occupant of the White House…The importance accorded by the media to this election is a strong indication of the essential role the US now plays on the planet, of their exclusive leadership (no matter how contested that leadership is, at least in discourse).
>
> (de Jouvenel 2004, 3)

Thus starts an editorial by Hugues de Jouvenel, Editor in Chief of the scholarly periodical *Futuribles*. The editorial proceeds to deliberate on the potential of Europe to offer an alternative vision of world order, one based not on "Anglo-Saxon free-market economics" but on the ideals of collective success, consensus and long-term concerns. This alternative appears less likely than ever to become reality, de Jouvenel believes, because "Today the planet is a veritable powder keg, characterized by the multiplication of risks of every kind," and as conflicts proliferate, Europe is held back by inefficient public institutions and excessive regulation. Europeans are preoccupied with economic stagnation while the real problems are in fact political. "It is therefore the political Europe we must construct, pursuing simultaneously the objectives too often opposed: deepening and broadening. We can thereby bring peace to our surroundings and promote democracy through dialog rather than trying to use force. There we have something that might confer originality on Europe in the eyes of the United States." (Jouvenel 2004, 4).

De Jouvenel is clearly expounding a Kantian perspective at odds with Kagan's view. We can identify in his arguments a constellation of themes – European power, "deepening" versus "broadening," American power, global disorder – that come together in contemporary French intellectual discourses. Rather than seeing the US as preserving peace and stability in the world, de Jouvenel sees the US as a force for disorder. This is the reason, hinted rather than stated outright, that Europe must move beyond its economic concerns to unify itself by addressing political problems. The means of doing so lie in deepening the political coordination and cooperation among European governments, currently included in the space of an open market. The objective of the next phase of European unification is not simply to make Europeans wealthier or more secure but to improve global stability by attaining the political power to effectively counteract US foreign policies. His editorial serves as an introduction to the *Europe-puissance* debate, one in which the future development of the United States is envisioned with trepidation and Europe provides a potential solution insofar as it can become deeper (that is, more politically and militarily coherent) rather than just becoming broader, that is, expanding in a geographical and demographic sense.

In the remainder of the chapter we consider the different angles taken by those who advocate a stronger Europe, a handful of opponents to such an agenda, and the representations of Europe and the United States that have motivated this call for a European power. The US is represented in these discussions at many points, not only as a force shaping the global order in which Europe must operate, but also as a threat to global stability and as a historical element that has made Europe what it is today. In its diverse roles we see not one "America" but several competing motifs, just as "Europe" is a constellation of motifs in European discourse – Kantian, Hobbesian, and in between.

We will focus on the strong Europe position and its opponents in turn. Those who support the project of *Europe-puissance* acknowledge the incoherent character of Europe as a political power and hope to clarify what this "unidentified political object" (Jacques Delors quoted in Bollaert 2001, 18) might become. Within the argument for a stronger Europe are at least six different positions taken singly or in combination:

1) Europe needs to be strengthened as a political body;
2) Europe needs a centralized executive branch including a president or presidents;
3) Europe needs a coherent and coordinated foreign policy;

4) Europe needs a social contract (a constitution and/or bill of rights);
5) Europe needs to control its own military defense rather than being tied to NATO, and therefore to the US;
6) Europe needs to promote an economic order that is an alternative to neoliberalism.

In short, the scope of discourse on a strengthened Europe includes economic, military, political and social elements. These arguments are not (yet) organized into a systematic socio-political-economic theory, but form a kind of hazy nebula. Any call to mobilize as Europeans also assumes the existence of competitors on the world scene which are the powerful non-European states – the US, Japan, and Russia – as well as two rising states that appear poised to assume a major role in the world system of the 21st century – China and India. European countries will be relegated to the sidelines of history, the reasoning goes, unless the EU (or some subset of the EU) organizes to form a confederation of nation states, perhaps even a republic. The integrated space of free trade that has been created under the auspices of the EU is therefore viewed not as an end but as a beginning.

Foremost among the intellectuals of statecraft arguing for a strengthened Europe is Hubert Védrine, who was the Minister of Foreign Affairs under Lionel Jospin from 1997 to 2002 and is now a prominent member of the Institut François Mitterand. As noted in chapter 1, Védrine's term *hyperpuissance* indicates the status of the US in the post-Cold War period, when its military, economic, cultural and political supremacy are no longer effectively contested by any other state. According to Védrine, the emergence of *l'hyperpuissance* as a new kind of geopolitical actor unprecedented in its dominance over the global system creates the need for Europeans to mobilize collectively and form a counterbalance that would be a means of restoring a multipolar world system. His advocacy of this project is persistent and highly visible, to the degree that one American observer complains of "Védrinism" (Caldwell 2000) in France. Yet while there does indeed exist broad support for Védrine's position (Poniatowski 2004, 153) Védrine himself rejects the use of his name as a banner, arguing that Védrinism "is too rigid a term. What is certain is that I've constantly tried to find a synthesis between the very rich heritage of French diplomacy … and adaptation to a context that has become totally different during the past ten years" (Védrine 2003c, 258). As for the precise nature of the new geopolitics Védrine makes very clear that it entails a shift from the multilateralism promoted by the US since the end of the Second World War to the unilateralism of the George W. Bush administration.

> We must be able to express differences of opinion without deserving an automatic avalanche of complaints from the Americans. One sometimes gets the impression that the smallest critique, even if it takes up comments formulated in the US by Americans, opens up a sort of badly arranged closet: everything falls on top of you and France is quickly called – by some at least – collaborator, Vichyist, liar, arrogant, tricky, mercenary, lazy, declining, etc. A series of Francophobic stereotypes assails you. It takes nothing more than that and soon enough certain specialized editors in Paris are denouncing anti-Americanism! It's exasperating! We absolutely do have the right, as old friends and reliable allies, to say what we think. As far as it concerns me, I have endeavored never to critique just for the pleasure of it.
>
> (Védrine 2003c, 261)

The first component of Védrine's geopolitical vision is the American *will to dominate* the world scene. "The Americans are certain that their political regime and their way of life are the best in the world. They can be messianic or Manichean. That does not make them colonialists or imperialists in the way the British or the French were" (Védrine 2003c, 14). A second component is American *disengagement* from forums and processes designed to permit collaborative decision-making among nations, provoking concerns regarding the unprecedented ability of the US to overrule the structures of collective decision-making that have governed international relations since World War II and that were developed, largely at the instigation of the US, throughout the twentieth century. The US foreign policy reversal came in spring and summer of 2001 with the refusal of the US to continue negotiations on the Kyoto Accord and the International Penal Court, the abandonment of the antiballistic missile treaty, and the abandonment of various other multilateral disarmament negotiations, all of which laid the groundwork for a preventive war prior to the catalyzing event of the terrorist attacks (Védrine 2003c, 13) and, combined with this detachment, an increasingly aggressive projection of nationalist interests through unilateral actions abroad. The implications of this change should not be underestimated: "When you think of all that has been proclaimed and hoped for over the past decades in the area of international law, of 'united' nations and of multilateralism, this is a revolution! What to do when confronted with this *hyperpuissance* which presents itself openly, without pretense or self-restraint? [*qui s'assume pleinement comme telle, sans fausse gêne ni managements?*]" (Védrine 2003c, 356). The third element of the new context is a fundamentalist movement in the US led by reactionary Christians opposed to fundamental tenets of modernity. US foreign policy now promotes a religious value system (ostensibly Christian) through military force while employing a religious rhetoric that masks self-serving motivations (see also Vernet 2002–2003, 804). The imbrication of politics and religion also leads to the emergence of an alliance between Washington and Israel's right-wing Likud party in the pursuit of closely interwoven anti-Palestinian, anti-Arab, and anti-Islamic agendas (Védrine 2003b, 4–5).

On this account, the US's version of the *mission civilisatrice* is perversely inverted; it is a forced march into the middle ages with their combination of armed conflict, servitude and religious fundamentalism. These developments combine with several deeply engrained features of American culture to render the US predominance in world affairs even more problematic. On the one hand, the American tradition of isolationism spawns a lack of concern about the rest of the world for its own sake but an eagerness to intervene whenever US security appears to be threatened. On the other hand, the foundational myth of the American nation pushing back the frontier between good and evil, order and chaos, is now transformed into US foreign policy for those elsewhere on the planet who do not share the American vision it becomes a problem (Védrine 2003b, 8). Thus if Védrinism exists it involves a belief that faced with a socially regressive hyperpower and its destabilizing tendencies, particularly the exacerbation of Middle Eastern conflict, the only effective means of regaining world stability is through the strengthening of European alliances so as to create a new global power that acts as a counterbalance.

The situation faced by the French people is not entirely unprecedented; its parallel can be found in antiquity as preserved in the powerful social memory – a kind of national mythology – of the Gallic tribes resisting the expansion of the Roman Empire: "Faced with this new Rome, what is needed is an ancient form of inflexibility" (Védrine 2003c, 357). This raises a second, and complementary, set of ideas advanced by Védrine: the need to counterbalance the US on the international scene. "But if influencing the United States is turning out to be almost impossible in the current conditions, then in the interest of global equilibrium and maybe even in the interest of the US, it is necessary to modify power relations and build a counterbalance at the international level and at the European level. A counterbalance is not necessarily hostile – consider the Euro – but it changes the game by setting limits" (Védrine 2003c, 17). This concern is founded on the idea that democratic political processes depend on a *balance of powers* – an idea central to American as well as French political theory at the domestic level where in both cases it promotes the functional separation of executive, legislative, and judicial branches of government, plus protection for journalism. But the merit of balancing powers is not self-evident to Americans looking at the relations *between* democracies because they cannot imagine their own country needing to be counterbalanced – they cannot imagine that global democratization might depend on the existence of a power capable of resisting US decisions or arresting US actions, because they equate US foreign policy with the preservation of democracy.[1]

Insofar as most European observers are less concerned with the American geopolitical evolution than he is, Védrine blames the European belief in the existing "international community" for this apathy. Europeans must be persuaded that force has once again gained the upper hand in the world (as during the rise of Nazism) and that there is essentially no likelihood of the US behaving as an equal among nations.

1 In this regard, Americans are more likely to assume that a single country can incarnate the best of the democratic spirit and consequently that that country should bear the burden of protecting democracy against the will of other democratic countries when those other democracies are unable to perceive the best course of action. The less perfect democracies must sometimes be overcome in international forums like the UN although such forums are sometimes useful as means of legitimating the decisions of the leading democracy.

Védrine therefore is as much a Hobbesian as a Kantian. Power politics are real, he believes, and that is precisely why international cooperation is needed.

In Védrine's eyes, the whole of the EU does not constitute a sufficiently coherent alliance to take on the role of checking US power, so what is called for is an alliance of the UK, France and Germany. This alliance, the *noyeau dur* (literally: hard core), could consist of a broader sub-group of the EU and possibly even Russia, but he prefers the most restricted core alliance for its strategic responsiveness (Védrine 2003b, 15). We will place this idea in context later in the chapter.

Védrine rejects the label of "anti-Americanism" for his stance, citing its "punctual" or specific quality as opposed to a blanket condemnation of all things American (Védrine 2004, 117). He also invokes the motif of global consensus: "In addition to the structural components of anti-Americanism, the Bush administration's style, marked by its blunt unilateralism, its opportunist ideas, its choice of vocabulary, and its war in Iraq, irritates the whole world with the exception of Israel and Kuwait" (Védrine 2004, 119). Like any form of chauvinism, blind anti-Americanism must be combated, he argues, but "the rise and fall of anti-Americanism in the future will depend on both the United States' use of its unequaled power and its ability to communicate with the international community" (in other words, hard power) (Védrine 2004, 119). How the US handles the Israeli-Palestinian conflict is especially important, but the fate of US strategy depends overall on whether the US "persistently acts unilaterally in conflicts [or] whether Washington accepts a dose of multi-polarity (French style) or even multilateralism (as hoped for globally)" (Védrine 2004, 120–21). So the US needs to be resisted not because of what it is but because of what it does.

The proscriptive element of Védrine's analysis, his call for European unity, assumes the need for checks and balances in the global political system akin to those built into the American political system. This idea, articulated in the term *contrepoids* (counterbalance) constitutes one of his most important contributions to the post-Cold War debate. The idea of a *hyperpuissance*, of the US as now operating from an unprecedented position of power, is the other principal contribution. Together, they pair an assessment of US power and a motivation for European unity locking the evolution of Europe and the US together in a single theoretical framework. This geopolitical framework resonates in the writings of many other French intellectuals and stands in direct opposition to the "hegemonic stability thesis" espoused by many Americans, which posits that global stability depends on a dominant state with superior values and a commitment to human welfare that graciously serves the world by preserving global order and stability (Dodds 2000, 21).

Marxist scholar Étienne Balibar uses Védrine's term *hyperpuissance* without attribution (2005, 17), demonstrating that the term is well on its way to becoming part of the French language as it spills into discussions that do not cite Védrine. Balibar argues that the US is now exercising its power to transform American exceptionalism into fact: to overcome trade protectionism of other countries while retaining its own protectionist policies, to judge other countries without being judged, to enforce global disarmament without having to disarm, to dissuade other countries from military actions without being subject to dissuasion, to control international political organizations (such as the UN) without being controlled by such organizations, and finally to inform the world without being informed by the world (2005, 134–41). Likewise the concept of *hyperpuissance* is filtering into the arguments of theorists who do not employ this precise term. Calvez (2004), for example, refers to the US as the "central superpower," and his exploration of the new kind of state supremacy wielded by the US owes much to Védrine's theoretical contributions. Alain Joxe (2004a) unleashes a scathing critique of the "Empire of Chaos" that is the world of today – a world under American domination but not under American control. Unlike a colonial power, the US is not interested in gaining territory or in civilizing the world, but prefers a world where barbaric violence chronically reigns in "zones of crisis" that are separated from the developed world by "institutional, economic or military membranes" and monitored from a safe distance by high-tech surveillance (Joxe 2004a, 17). The *Védrinistes* are uncomfortable, above all, with the existence of a world system dominated by a single-superpower, completely apart from questions about how well (or poorly) that power may be governed under a particular administration. As political theorists in other countries begin to draw on the French geopolitical discussion to enrich their theories, the concept of *hyperpuissance* is likely to find echoes in other languages.

What sets Védrine apart from most of his followers is his sensitivity to dynamics of the American political process that are normally overlooked abroad. He recognizes, for example, the foreign policy shift that occurred between the Clinton administration and the George W. Bush administration, and notes that many of the strongest indications of *hyperpuissance* are grounded in the policies initiated under the Bush administration. Other French observers tend to overlook or deliberately obscure the power struggles within the US, but Védrine is well aware of the contradictory forces in the US that give rise to a range of foreign policy approaches. He even voices appreciation for Bill Clinton's approach to the Middle East and complains in relation to the present administration that "we have never seen an American administration consult so little with allies, or even, in some cases, treat them with such complete disregard (*disinvolture*)" (Védrine 2003c, 11, 262). Nonetheless he does not advocate simply waiting out the Bush administration.

> Ultimately, as much as the reactionary Southern fundamentalism of the Christian coalition and the macho hubris (*hubris botté*) of the Pentagon appear specific to this administration, it is equally true that unilateralism, tempered slightly with permissive multilateralism on the side, would appear to remain necessarily an enduring element of American policy.
> (Védrine 2003c, 356)

Variations on Europe-puissance

Védrine's successor at the Ministry of Foreign Affairs (and subsequently the Prime Minister) Dominique de Villepin (2004), takes up the call for European unity in *Le Requin et la Mouette* (The Shark and the Seagull) – the two animals that represent the complementary facets of geopolitical power (essentially Nye's soft power and hard power). De Villepin echoes Védrine's call for a counterbalance to US power and his perception of American power as out of control (while adding some rhetorical flourishes).

> The United States, fatherland of Wilsonian idealism and the United Nations, has recently opted for unilateralism, the other pole of their diplomatic doctrine. The only great power capable of exercising their domination at the scale of the planet, the United States maintains an incontestable superiority in strategic, technological and military domains; their territory has never been subject to a major menace even if September 11 revealed a certain vulnerability; their culture penetrates the rest of the planet; the American enterprises are the principal beneficiaries of globalization which gives the country a strong growth. Finally, American domination is based on the assertion of a certain number of values and principles.
>
> (de Villepin 2004, 57)

In choosing to respond to problems through hard power the US has left an opening for Europe that is both a need and an opportunity: "When the weak can make the strong falter, when ideologies scoff at the most elementary rights, falling back on force is not a sufficient response. More than arms, what we need today, in order to create order and release tension, is a conscience and an example" (de Villepin 2004, 62). In effect, Europe's soft power is needed because of the adverse effects of modernization and secularization. "Our modern societies appear to want to construct themselves around a center that cannot be found because shared values are missing" (de Villepin 2004, 40). European soft power is also needed because lack of understanding is increasing despite, or perhaps because of, modern communications: "Our societies observe each other every day through interposing screens, but suspicion remains nevertheless" (de Villepin 2004, 63). As the world's "public conscience," Europe can defend the ideals of international law and cooperation, ideals the US promoted throughout the 20th century but ultimately rejected with the anti-historicism of the first George W. Bush administration and its dependence on hard power. In the following passage de Villepin articulates these ideas:

> But more than a model [of democracy] what Europe brings to the world is a conscience. A conscience that is not, as some claim, a clean and satisfied conscience, pretentious and arrogant; rather it is an uneasy conscience, preyed on by doubt, always in the search of an elsewhere. Europe has recently discovered the magic of creative differences, the virtues of the confrontation of cultures and the meaning of dialog. But we have not forgotten that the landmarks of our history are conquests and acts of oppression. … Our legacy to the world is the consciousness of time, the sense of history and the concept of progress, but we also have breathed life into this anthropocentric demon that has devastated the planet, this attitude of domination, this demonic hubris that reduced people to slavery, destroyed cultures and trampled difference underfoot.
>
> (de Villepin 2004, 243–4)

So Europe's colonial past is a resource rather than a handicap. The guilty consciences of former empires offer the world the ideological justifications for cultural diversity and a check on future imperial impulses. On this account the European sense of historicity – deep, complex, and troubled – supports not just soft power to attain Europe's goals, but also a kind of reflective moral leadership that the US is unable or unwilling to offer. De Villepin waxes poetic on behalf of Europe but he has no qualms about engaging in nationalistic puffery:

> France, mid-range power, country like all the others? No, rather a power in the service of peoples, an awaited power, hoped for and listened to, in love with the values of tolerance, democracy and peace. Power at the service of a Universal it has had to defend throughout the centuries with weapons and with laws, a Universal that cannot live today without respect and exchange.
>
> (de Villepin 2004, 9–10)

Others echo these sentiments. De Charentay asks, "How can the dialogue between America and Europe be serene when we are on such different paths: one, of self-affirmation, the other, of multiple questioning? More than different policies, what separates us are different cultures: one of dreams, projection, and action, faced by one of questioning, doubt, introspection and discussion, with all the hesitations that follow" (2005, 8). To think of social memory as a *resource* is fundamentally alien to the American world view in which doubt is perceived as negative because it interferes with action. In contrast, American pragmatism predisposes Americans to doubt doubt itself. But Charentay argues (2005, 8): "The American dream remains fragile, founded as it is on commercial debt, military power, and extreme individualism," while "in refusing the path of power [the Europeans] probably take the more sustainable route." If an action-oriented US has less and less in common with a reflection-oriented Europe, then the French may need to help develop a stronger Europe as a space in which to build alliances that will facilitate not just reflection but also action.

A primary voice on the left is that of Emmanuel Todd, a historian and demographer who argues that the United States has entered a period of decline and it is time for the "emancipation" of Europe from American control. Germany, the UK and France should take the lead in rejecting the US's self-serving "leadership" in Europe, while Japan should do the same in Asia. Although he touches frequently on military and social issues, Todd's best known argument is that the US economy – an "omnivorous universal consumer" or, more graphically, a "black hole" (2002, 258, 175) – can provide only the American dollar in exchange for material products from around the world, and the dollar's value is in turn supported through transfusions of foreign investment capital, coercive trade policies and military intervention. "God definitely does not bless the America of today," Todd declares, "this is a country that denounces evil everywhere because it is itself turning evil. Its regression can make us aware of what we are in the process of losing: the America of 1950–65, the country of mass democracy, of freedom of expression, of expanding social rights, and of the fight for civil rights was the empire of good" (Todd 2002, 172, 286). But what remains of the nation is useless: "The world has no need for this America: militaristic, agitated, uncertain, anxious, projecting its internal disorder onto the planet" (Todd 2002, 286). However much Todd may idealize the America of the 1950s and 60s, he pins his hopes for the future on an entity he refers to as the "United Nations of Europe" led by a "directing triumvirate": Germany, France, and the UK, engaged in a solid alliance with Russia and Japan.

Karoline Postel-Vinay sees Europe's offering as a particular concept of globalization that serves as an alternative to the bipolar freedom vs. terrorism model dominating US policy and rhetoric. "It is possible that, aided by the political maturity of the Union, the Europeans will be able to value in some small way their own experience of regional cooperation, and from there may elaborate a different geopolitical narrative [from the Americans] for the 21st century" (Postel-Vinay 2005, 26). So the promotion of a geopolitical vision – one based on process rather than pattern, on multiplicity rather than binary division – would be Europe's contribution to the rest of the world.

Axel Poniatowski, representative to the French National Assembly from the Val d'Oise and president of the France-United States Friendship Committee in the French National Assembly, argues that the US and France are on divergent paths largely *by default:* Americans have become unilateralists because the US lacks real competitors, while the French have become multilateralists because of their loss of power over the past century and the consequent need to form collaborative ties in order to regain that power (Poniatowski 2004, 15–16). His argument is similar to that of Robert Kagan (2002) in terms of the link it draws between political culture and geopolitical power, but he draws very different conclusions from this link. The US has suffered a profound shock with September 11, and the emotional "scar" *is the sense of being at war.* Meanwhile, the French have rejected the idea of war as the definition of the world situation, despite supporting the US invasion of Afghanistan. So the US's continued pursuit of the "war on terrorism" in Iraq and elsewhere has driven a wedge between the two countries. He adopts Nye's division between hard and soft power operating in the global arena and allocates hard power to the US and soft power to Europe. Poniatowski portrays the US as manipulating the European Union over many decades to deliberately undermine Europe's power and thereby control a potential competitor. He voices concern regarding countries like Spain, Poland and Italy that suffer from a "mid-level power complex" that drives them to curry the favor of the US, an "orphan syndrome" (Poniatowski 2004, 153) that causes mid-level and small European states to turn to the US as the strongest protector, in lieu of a *Europe Puissance* that does not yet exist. What is therefore needed for the sake of Europe and the world is a geographically limited, autonomous European bloc led by France. "Between France which wishes for a politically autonomous and militarily powerful Europe, England which is having trouble cutting the umbilical cord with the US, and Germany which remains pacifist, always dreading its old demons, the synthesis is complicated. France has the job of directing this evolution" (Poniatowski 2004, 212).

For Daniel Vernet (2002–2003) Europe's internal debate boils down to a dispute between *eurosceptiques* who see EU communiqués as so many empty words versus *europhiles* who believe European alliances will help European states preserve some measure of sovereignty in the face of globalization. "Neither France, nor Britain, nor Germany can any longer, on their own, aspire to argue with the United States for a multipolar world. To share their efforts they must consent to new transfers of sovereignty, as they have already done in the commercial and monetary domains. They struggle, the first two in particular, to get over the fear of a loss of identity. They know however, at least since the calamitous Suez raid in 1956, that alone they no longer matter" (Vernet 2002–2003, 804). The future of Europe is defined by the American threat – not a direct military threat, but a threat to sovereignty and control over foreign affairs, demonstrated over decades. Vernet directly engages Kagan's accusation that Europe is living in the "Kantian" delusion of a post-historic paradise by attacking Kagan's argument on three counts: factual inaccuracy (Europeans have applied the Hobbesian worldview when appropriate), excessive speculation (no one can say with assurance what the world would be like if the US had not played the role of global policeman), and reductionism (the world does not divide simply into good and evil).

Kagan's initial conclusion, that Americans and Europeans live in "different worlds" as a consequence of the past 50 year period in which the US has played the role of policeman to the world while Europeans have enjoyed the effects of this pax Americana, is nonetheless widely accepted by French authors (Le Guelte 2003; Duchêne 2002–2003; Vernet 2002–2003; Parmentier 2002–2003; Fontaine 2002–2003). The premise that playing a particular geopolitical role, be it protector or protected, will affect political culture, is not disputed. What is disputed, however, is whether the US is wiser or better for having internalized the role of the superpower (and later hyperpower), for living in a Hobbesian universe, for playing the role of policeman, and/or for putting itself above the law, and specifically whether the world can be made a better place by a "leader" like this. In the eloquent words of Robert Toulemon:

> Europe, certainly not yet finished, is the entering into political reality of the grandiose Kantian concept of a republic of nations. It is also the rejection, after the catastrophes of the early 20th century, of the Hobbesian concept of rivalry between sovereign states. Henceforth the survival of all of humanity is conditional on the construction of a universal democratic order. This responds as much to the demands of the coldest realism as to the highest ideological aspirations. That should be our response to those who, like the American, Kagan, accuse the Europeans of being converts to pacifism just because they lack the means to power.
>
> (Toulemon 2005, 98)

On this account, Europe is not coalescing as a power for its own sake but in order to protect global peace and stability. George Le Guelte argues along a similar line:

> If other countries continue to present themselves as indifferent to military engagements [of the US with regard to] the establishment of social rights, the outline of future international relations is that of a world where not a single rule is respected, where blackmail becomes a habit, where the number of countries with collapsing social structures mounts, and where criminal organizations develop their activities without restrictions – a situation of chaos punctuated from time to time by arbitrary preemptive strikes from the US, which can do nothing but aggravate the tensions.
>
> (Le Guelte 2003, 26)

This inspires a call for a European role that is active, but not in the same way the US is active. Europe must act in the ideological realm, as a Kantian participant in the world order. "[T]he developing transatlantic ensemble has no need for Europe to add new tanks or bombers to an American arsenal already beyond numerous. The most useful contribution of the Europeans will be to furnish a coherent alternative to the doctrine of the neo-conservatives" (Le Guelte 2003, 26).

Similarly, Adrien (2001) argues that "An interesting Europe for the world and for the US first off, is not a weak Europe that follows the echo of the voice of whomever, but a Europe endowed with ideas and the means [of achieving them]." This call for ideas, for alternative ideologies, is not limited to one or two French intellectuals. It resonates throughout French scholarly debates about the future, about world politics, and about Europe, France, and the US. But ideological power is a nebulous concept. Clarifying its ingredients remains a formidable challenge.

For some, European power would require a centralized executive body including a European president or presidents (Védrine 2003; Saint-Etienne 2003, 139–42). For others, the fundamental need is for a coherent and coordinated European foreign policy (Duchêne 2001, 778). Some call for a social contract in the form of a European constitution and/or bill of rights (Saint-Etienne 2003, 166–69) while others urge an alternative mode of development to replace economic neo-liberalism (Le Guelte 2003, 26). Still others argue that Europe must make the investment required to build a significant defense force and must cease to depend on US military protection (Poniatowski 2004; Todd 2002, 244; Saint-Etienne 2003, 179; Colombani 2002, 81).

In short, a new basis for European power may be seen to reside in authorities or documents, diplomacy, economic leadership, or armament. Opposition to the United States is, however, a recurring element because US power dominates individual European countries in so many different areas: economic, political, and military. Theoretical differences among the advocates of a stronger, more unified Europe have not led to major contention at this point in the *Europe-puissance* debate because of the supremacy of the US in so many different areas. The US's strength generates a kind of consensus because even when the terms of the critique are not the same the subject of critique is clearly the US and the problem, American power. This constitutes what I have called the *Achilles heel* of the global hyperpower. Hard power is a distinct liability when it becomes so overwhelming that it actually generates soft power for its opponents as soon as they engage in critiques or in acts of resistance.

Atlanticists and nationalists

Still, the US is not without a measure of soft power, even in France. Some French observers refuse to join in the rhetorical construction of the US as the "Disruptive Other" that has justified the *Europe puissance* project. Such observers hold a line that resonates with Washington's Hobbesian view of the world, while others are more worried about Europe as a threat to French sovereignty than about the US in the same capacity. This group of observers is therefore divided into Atlanticist and nationalist segments. The Atlanticist position fully endorses the Euro-American force embodied in NATO and defends the US-Europe alliance as a long-standing force for democracy, peace, prosperity and stability in the world. This stance is often justified by invoking economic constraints (Europe cannot afford to become a military rival to the US) and adopting American motifs of the US fighting fascists, communists, terrorists, and other enemies of freedom. This view of history is associated with the extreme Right in France. The nationalist position, in contrast, places unlimited faith in "the hexagon" and sees other states primarily as rivals or threats to France. On this account European unity would destroy France. Atlanticists and nationalists have little in common aside from their inherent conservatism and disinterest in strengthening Europe. Moving deeper into these positions, however, several distinct positions can be identified in the Atlanticist corner of the debate:

1) Europe and the US are so closely allied that discussion of *Europe-puissance* constitutes separation from a binding and essential political alliance;
2) Europe would conduct its foreign policy much like the United States if it were in a similar situation, so the US should not be blamed for its resort to the military option;
3) Europe should not spend money on military defense because the US already ensures global security, while Europe plays a different role in the global system by promoting peace, cooperation and diversity;
4) Terrorism is the main threat in the world today and Europe and the US cannot meet the challenge of terrorism separately.

Laurent Cohen-Tanugi is one the most prolific Atlanticists. In *Les Sentinelles de la liberté: L'Europe et l'Amérique au seuil du XXIe siècle* (*The Sentinels of Liberty: Europe and America at the Threshold of the 21st Century*) he calls for a thaw in US-Europe relations while arguing at the same time that the rift between the two continents has been overdrawn by the press and the popular media. The US is, in his view, the leading guarantor of world order, peace, and democracy. He pleads for a "new alliance" across the Atlantic and an end to the "ricochet" of hostile attitudes (2003, 194) (precisely what I have tried to capture in the term "reverberation"). He accuses Europeans of employing anti-Americanism as the "cement" holding Europe together, declaring that "the principal danger resides, in truth, in the insidious media-political indoctrination (*l'insidieux endoctrinement politico-médiatique*) of public opinions in favor of the separatist thesis" (2003, 199, 201).

This attempt to brand European sovereignty as "separatism" arises out of a passion for the United States, as articulated in his earlier writings which idealize the US's free-market economy as "the temple of socio-economic liberalism" and scathingly critique the French government for its "brutal and artificial invasion" into the private affairs of its citizens and reducing market forces to no more than a "residue" (Cohen-Tanugi 1985, 26, 131). For Cohen-Tanugi the state has no business directing the national economy or redistributing wealth. Even this staunchest of French Atlanticists cannot, however, help but mention "the systematic and brutal rejection of many international treaties in the global interest by the Bush administration – from the Kyoto protocol to the International Tribunal, including treaties on landmines and light arms sales" (Cohen-Tanugi 2003, 55). But he excuses this behavior on the premise that Americans needed to reject these treaties in order to effectively maintain peace and prosperity throughout the world. He places the blame instead on Europe's lack of diplomacy (Cohen-Tanugi 2003, 210). Likewise he mitigates concern about the US's "pre-emptive strike" on Iraq, judging that strategy to be "less a revolution in the history of international relations than an application of good sense in the present context" and he assumes the mantle of apologist for the US: "a Europe in the position of a superpower and the primary target of Islamic hyperterrorism would adopt similar positions on these questions to those of the US" (Cohen-Tanugi 2003, 211). In the context of his celebration of American free-market values and his apologetics, he calls for an *approfondissement* (deepening) of Europe, but his vision largely matches that of American conservatives who see the EU as a force for the diffusion of order and democracy through the free market and lowered trade barriers counterbalancing or countering the US's self-assumed international policing mission (1992; 1995).

Others oppose *Europe-puissance* while condemning the US. One type of position that does so is a realist position based on viewing the global order as a competition of inevitably unequal forces. On this account different levels of power dictate that countries should conceive of their roles in the global theater in different ways. Philippe Raynaud, a professor of political science at the École des Hautes Études en Sciences Sociales (EHESS, Paris II), argues that the US cannot reasonably be challenged or equaled in its areas of strength, even if it is deeply flawed. He argues therefore that Europe and the US must play different, equally necessary roles in the world. A kind of division of labor, with the US having "prerogatives that are difficult to universalize," such as nuclear deterrence, while Europe, pushed by its sense of rivalry with the US, works in a different direction, assuming the responsibility to propound the moral and legal foundations of democratic society. "From this point of view," he writes, "the valorization of the European social model as well as the denunciation of the death penalty, far from being 'gimmicks,' play a major role in the construction of a European alternative social order; it remains to be seen if they can suffice to compensate for the relative weakness of Europe, relative to 'classic' sources of power" (Raynaud 2001, 614). *Europe-puissance* is, on this account, the wrong idea of European power because of its military implications although Europe does have a special role. The idea of a symmetrical counterbalance to the US is an illusion and Europe's avocation is instead to offer the world a more "soft and humanitarian" (i.e. Kantian) version of democracy than demonstrated by the US, and to "enthusiastically critique the primitiveness of their morals and the harshness of their laws" even while turning to them to collaborate on answers to the most difficult international problems (Raynaud 2001, 615–16). With regard to the future, a United States of Europe is not an appropriate model; the EU would be better conceived as "a sort of Canada with a temperate climate" (Raynaud 2001, 615) which is to say a decentralized federal republic embracing diverse nations.

In a very similar vein, François Duchêne (one of the principal collaborators with Jean Monnet on the European Coal and Steel Community) believes the EU has no need for the military strength it lacks. Military weakness in fact can be an asset in the present situation where political strength matters more. Whereas the United States' massive investment in armament has not brought it security, it has permitted Europe to focus on economic development to the point that "the European Union can speak with the United states as one equal to another" (Duchêne 2001, 607). In the political realm the EU is a match for the US because its stances are based solidly on European public opinion which cuts across borders (Duchêne 2001, 608). Fortuitously, Europe's strengths are those best suited to the state of the world today. "It is therefore in the most promising domains on the international map that the EU is best equipped for action" (Duchêne 2001, 607).

Jean-Marie Colombani, the editor of *le Monde*, is another prominent Atlanticist. Columbani penned an infamous front page editorial that appeared on September 12, 2001 under the headline, "Nous sommes tous Américains!" (We are all Americans!). His declaration of support for the US aroused a firestorm of controversy and his book-length essay, *Tous Américains?* (Colombani 2002), if not an apology or a justification, it was at least an acknowledgement of this controversy. In his view, the US may be misguided, particularly under the leadership of George W. Bush and the neoconservatives, but the global struggle against terrorism, particularly against Islamic fundamentalism, is vitally important and requires the concerted action of the US and Europe. Colombani parts ways with those who cast the US in the role of threat to global peace, declaring that the greatest threat to world peace is in effect a nascent fascism in the Muslim world, and asking "is an impoverished fascism any less fascist?" (Colombani 2002, 144).

In contrast, Jacques Delors, the former president of the European Commission and president of the pro-European think tank "Notre Europe," frames Europe's sphere of concern as regional rather than global. His argument is that efforts should be focused on developing cooperation and dialogue within the framework of the EU and improving relations on Europe's southern and eastern margins. He calls for a European political structure lying between familiar categories, "To return to terms that are familiar to some, … more than a European space and less than a *Europe-puissance*" (Delors 2003, 11). This middle road is, unfortunately, difficult to articulate and depends on hazy terms like "active peace" and "management/valorization of cultural and ethnic diversity." On this account, Europe must manage Northern Africa and Eastern Europe (as they pose direct threats) but not the entire world.

Various supranational organizations are likely to become contested battlefields in the struggle between Atlanticists and nationalists, most notably the UN, NATO and OECD. Calls for reassessment of the French role in these organizations have circulated in public affairs journals as well as in popular media. For example, a well-known Jesuit scholar joins the Atlanticists and argues that the best short-term solution to the monopolization of power by the United States in NATO and the UN is to officially recognize and in fact institutionalize that "central superpower" status (Calvez 2004). He believes that playing along with the fiction that these organizations are composed of equals simply increases the legitimacy of US power and disguises the unilateral nature of the current decision-making process; conversely, if the US were to be opposed outright, by France or even by Europe as a whole, it would simply find ways to "go around" as it did with the decision to invade Iraq, and Europe would find itself even more thoroughly marginalized. Calvez's solution (2004, 27–30) is to more openly acknowledge the "asymmetry" of international organizations by setting up a formal consultative body including non-Americans to advise American policy-makers, by expanding the membership of the UN Security Council to include permanent seats for all of the current world powers, and by permitting the UN Security Council to operate on "consensus minus one" rather than the current consensus basis which would allow a unilateral US position to be overruled. His suggestion is bracketed by the admission that care must be taken that "recognition" of the US as the "central superpower" not lead to the kind of appeasement that Germany enjoyed prior to World War II. So in international decision-making bodies the US may find itself acknowledged as the leader and challenged at the same time.

Among the opponents of *Europe-puissance* are also, of course, a certain number of nationalists. French nationalism may be marginalized in the intellectual community and the mass media, but the rejection of the European Constitution by the French voters in 2005 demonstrated a persistent strain of nationalism (allied with socialism). Hervé Beaudin's *Au-delà du "non"* (2005) demonstrates a confluence of xenophobia, nationalism, and Marxism that must certainly cause consternation among those who support *Europe-puissance.* For Beaudin, "the victory of the "yes" would have once and for all hitched [France] up to a Europe functioning very well without us and constructed in opposition to us" (2005, 9). This Europe, in Beaudin's view, is an "antidemocratic" organization subjecting the French people to an "anglo-American" cultural and legal system, and imposing a kind of "corporate feudalism" (2005, 18–19). Likewise the idea of joining forces with other European countries in a European Security and Defense Policy (ESDP) is deeply flawed, in Beaudin's view, because such a body would necessarily buttress the functions of NATO, unless explicitly divorced from NATO, and therefore the ESDP is a de facto tool of US interests. Beaudin is convinced that the US already controls not only the other European countries but the entire process of European unification. Only by retracting obligations to Europe can the welfare of the French people be safeguarded from the tentacular influence of the US.

The Europe-puissance position, in contrast to these Atlanticist and nationalist positions is much more pervasive. In the circles of scholars, policy-makers, journalists and other "intellectuals of statecraft" the position has many proponents, and their arguments are eloquent, persuasive, and ubiquitous. In the words of one observer, "Europe exists a bit more every day on the international scene, and it plays a favorable role there. But the reality is that for a long time already there have been shared [European] positions, in increasingly numerous cases, coexisting with distinct national desires on major subjects" (Adrien 2001, 315). But as a political agenda *Europe puissance* is faced by nationalism, localism, socialism, and other political forces reflecting different views of community, and still capable of political mobilization. Adding to the tension between Europeanists and Atlanticists is disagreement on the meaning of "Europe," particularly around the relationship between a core (invariably including France) and a periphery that is defined in a plethora of different ways.

Bounding Europe

French geopolitical discourses in intellectual circles are unresolved regarding the scale of European political organization that would be best suited to respond to external and internal threats to Europe's security, prosperity and stability. For most observers the emerging political entity takes a dual form: the larger economic space of the EU (now referred to as the EU-25 in reference to its 25 member states) but also a politically united subgroup of these states that would have the ability to react in a quick and concerted way in the international arena through diplomacy and/or military intervention. This translates into the geographical idea of a European core and periphery, and in the historiographic motif *Europe de deux vitesses* (Europe of two speeds). Together these motifs imply that the periphery is not locked out by the unification of the core, but its incorporation may need to be delayed. According to various authors, the European core would consist of:

1) a two-way alliance between France and Germany;
2) a three-way alliance of France, Germany and the UK;
3) the EU members from the pre-2004 period, the "EU-15";
4) anything but the current incorporation of Europe as an American satellite region.

Daniel Cohn-Bendit, a representative in the European Parliament and political figure in both France and Germany, articulates the new dream of Franco-German leadership: "It is no longer a matter of the responsibility of France and Germany to reconcile two peoples, but the reconciliation of France and Germany for the [sake of leading the] world" (Cohn-Bendit 2005, 82). Bernard Adrien (2001) also voices support for the *couple franco-allemande* arguing that these two countries have invested more in Europe "for evident historical reasons" than any other countries, and they should invest still more because it is their only means of "regaining an influence in the world proportional to their combined weight" (Adrien 2001, 317). Adrien argues, in addition, that the bargaining involved in building Europe has in effect marginalized France and Germany from what should be their rightful position as European leaders. Henri Froment-Meurice (2001, 15) invokes the French-German pair as the key to resisting US dominance of Europe through NATO, yet mentions as well the possibility of a three-country European core, mocking American fears that Germany and the UK will be affected by "the Gaullist poison." Emmanuel Todd's vision of the Europe that will take the US's place as the leader of the world is a rather vague entity: he refers to the "couple leader germano-français" but also invokes the "directing triumvirate" of France, Germany and the UK, and even speculates optimistically about a global alliance between Europe, Russia and Japan: "The ensemble of Europe, Russia and Japan represents more than two and a half times the American power. The strange activism of the United States in the Islamic world ceaselessly pushes the three northern powers in the direction of a long term rapprochement" (Todd 2002, 86, 275). Todd's geographically ambiguous view reveals the way in which almost any formation of power can be seen as preferable to the existing unilateral dominance of the world by the US. This emerging core of European and global order would be able to neutralize the disorder sown by the United States because of its intrinsic understanding of the principles of "secularism, peace and equilibrium" which, he argues, are completely alien to Americans (Todd 2002, 246). Within Europe, the most problematic of the potential core countries is the UK, because of its close ties with the US, so Todd pins his hopes above all on the "leading couple" of Germany and France (Todd 2002, 275). This narrowing of European ambit also serves to reconcile his interest in creating a counterbalance to the US with his Gaullist impulses toward economic protectionism (see Todd 1998). In his view, a French-German union could presumably pursue a policy of import substitution and reject the misguided policy of free-trade, as well as presenting an obstacle to US strong-arm tactics that force countries to sacrifice national interests in order to feed the US's voracious appetite.

Echoing the call for a two-way alliance, the Assistant Director of Institut Français des Relations Internationales (IFRI), Dominique Moïsi, points out that while the US once acted as the matchmaker for France and Germany, promoting their cooperation after World War II and later fighting to expand this alliance to other European countries, today the US's primary attitude toward both countries (and Europe in general) is one of indifference (1995, 61). Recalling furthermore that in 1963 the US undermined provisions of the Treaty of the Elysée forcing a broadening of the Franco-German alliance, he then indicates that current US disinterest in Europe may allow resuscitation of the Franco-German "marriage." In his view, any future strengthening of the alliance would have to institutionalize a framework balancing the power of the two countries while finding ways to harmonize their different views of Europe. "In Paris as in Bonn there is the desire to work with the other capital to create Europe, but it is not the same Europe. The Europe seen by France is always more intergovernmental and diverges from the federal Europe conceived by the Germans" (Moïsi 1995, 68).

In *La puissance ou la mort* (power or death) Christian Saint-Étienne (2003), an economist with experience at the IMF and OECD, argues that the EU is currently too uncoordinated to affect world affairs despite its great mass. "Mass" in this case refers to numbers of people and geographical extent, neither of which increases Europe's power in his view. The Europeans have been reduced to the status of "critical spectators of a world that escapes them" (Saint-Étienne 2003, 29) and unless they restructure their union, history will be shaped by the interplay of the United States, Russia, China, India and Japan. The solution is not to add on to the already swollen ranks of EU member states, but rather to construct a new European coalition with the unity and coherence that would enable it to again play a prominent role in the world. "If we desire to move beyond this peculiar moment in history, it is appropriate to affirm that for the European countries that are so inclined, it is necessary to 'decide to choose' in order to exist: to choose between population mass and power, between apathy and action, between existential renunciation and the political and philosophical project of Europe" (Saint-Étienne 2003, 165).

Saint-Étienne therefore proposes strengthening some portion of Europe, making a "united Europe" within the "open Europe" of the free-trade zone (Saint-Étienne 2003, 73). He is most enthusiastic about Franco-German unification for which he suggests the term *République du Rhin* (Republic of the Rhine). Germany earns the status of France's special partner through its strength, its demonstrated autonomy from the United States, and its intellectual traditions: "France and Germany share the heritage of enlightenment philosophy which permits them to establish a secular democratic republic, reason and justice serving as foundations for public decision making" (Saint-Étienne 2003, 179). The already strongly integrated Benelux triad might also be admitted to this Republic (at any rate they would not greatly affect its power or coherence) and he admits that a coherent political space could occur within a union of up to twelve European member states. From his arguments it is clear he believes the new member states added in 2004 are little more than dead weight on Europe, and the plans to incorporate Cyprus and Turkey amount to an "existential trap" into which the US is trying to force Europe (2003, 83). Saint-Étienne's Republic of the Rhine would offer what he calls "sustainable modernity" as well as "sustainable positive power" (Saint-Étienne 2003, 168) by separating state power (military and police) from politics, recognizing the rights of individuals, prioritizing the rule of law, and maintaining free elections. "Positive power" consists of exerting global influence (soft power in Nye's terminology) on behalf of long-term global objectives serving just causes (Saint-Étienne 2003, 20). Although terse in his references to the US, it is clear that Saint-Étienne believes the US has become a threat to global stability and the most nefarious aspect of the US is its pressure for the incorporation of Turkey in the EU which would, he argues, make the EU incoherent by expanding its borders to Soviet Georgia, Armenia, Iran, Iraq, and Syria. Similarly, Henri Froment-Meurise, the former French ambassador to Moscow and Bonn, and author of *Un Puissance nommée Europe* (*A Power Called Europe*) responds to Zbigniew Brzezinski's call for Europe to include the Ukraine, Georgia, Azerbaijan, Turkey and Israel: "without entering into a discussion here of such an extension, it is clear that it would bury still deeper the entire project of a *Europe-puissance*" (Froment-Meurice 2001, 14).

The call for *Europe-puissance* echoes among various other French political theorists and essayists including Laurent Cohen-Tanugi, Axel Poniatowski, and Daniel Vernet, each of whom argues that a stronger Europe does not imply a larger Europe. Indeed, while the US has been eager to promote Turkey's incorporation into the EU (under the assumption that it will accelerate Turkey's transformation toward a more democratic society and promote the diffusion of democratization into the Islamic world), these authors fear that the EU has expanded too much for its own good and for the cause of global stability, and that what is needed now is more coherence. Whereas US policy-makers and much of the American public adopt a civilization model of the current global order (articulated for example by Samuel Huntington 1998), and therefore see the inclusion of Turkey or any other country in the EU as an additional entry in the list of states that are "with us" rather than "against us," potential additions have quite a different significance when viewed through the *Europe-puissance* lens.

Arguments like these about the problems of incorporating Turkey may be justified, but they inevitably garner the support of French racists and religious chauvinists. The concepts of *noyeau dur* and *Europe de deux vitesses* therefore reaffirm right wing sentiments but also receive support from the left as means of counter-balancing the US-led neo-liberal world order. Such views of course stand in direct opposition to the view of American political theorists who see the US as specially endowed with the right and responsibility of unilateral military actions for the sake of humanity (Rubenfeld 2004).

Defining the United States

If, judging by these arguments, Europe occupies an indeterminate space on the earth surface anchored by an uncertain set of core countries, the US is no more clearly defined. Aside from disputes about the moral and intellectual value of US contributions to the world, serious differences of opinion arise among French observers with regard to the US (though in this case disputing its power rather than its boundaries). America is quite multifariously seen as:

1) a wholly new kind of global power, a dangerous "hyperpuissance" defined by its overwhelming power and lack of challengers;
2) an unrivaled bastion of freedom and democracy;
3) a tottering giant nearing collapse due to its economic decline and cultural regression;
4) a strong country but one that is increasingly rivaled by and increasingly dependent on other populous or wealthy states such as Japan, India and China.

Védrine's concept of *hyperpuissance* has been introduced earlier. To add to our understanding of the term we must recall that the US was already a superpower when it faced the Soviet Union. The fall of the Soviet Union propelled the US into a new geopolitical situation requiring a special term for the new geopolitical perceptions, conceptions and representations. Now the global system takes the form of a hub and spokes, with the US as the hub and all the other states as the spokes (Védrine 2003a, 2003b).

The most striking contrast in French intellectual opinion with regard to the US exists between Védrine's *hyperpuissance* and Emmanuel Todd's vision of the US as a giant teetering on the brink of collapse. Todd's *Après l'Empire* (2002) was tremendously popular in France following the attacks of September 11. Having successfully predicted the fall of the Soviet Empire, he made a similar prediction with regard to the US and applied a historical methodology to detect ominous signs in the United States: economic parasitism indicated by falling industrial productivity and a rising trade deficit, a weak and ineffective military reduced to theatrical displays against the weakest countries of the world, a primitive form of society in which universalistic ideals have not yet been separated from racist attitudes, growing segregation, indiscriminate application of the death penalty, support of blatantly unjust regimes abroad (particularly Israel), and finally a statistic Todd finds particularly revealing, an increasing infant mortality rate. Furthermore, the US is trying to stabilize an unbalanced global economy in which it is the prime beneficiary but it lacks the military or ideological means to do so (Todd 2002, 177). In sum, the US is suffering from "imperial overstretch" (a term he adopts from Paul Kennedy) combined with internal decomposition. In face of these manifest indications of decline, Europe's challenge is not to keep up with the US but rather to protect itself against the disruptive effects of the tottering giant's imminent collapse. The success of Todd's book is indicated by its re-issue in the inexpensive "folio" format, which suggests in turn that the habit of dismissing the US as backwards and primitive, which Roger traces to the colonial period, (see chapter 2) still holds a strong attraction in France today. Todd derives considerable authority from his successful prediction of the fall of the Soviet Union based on economic and demographic indicators (1976), but his popularity has not led many other opinion leaders in France to echo his pronouncements of imminent American decline (see Gréau 2003, 33). One who does follow Todd's lead is André Fontaine (2002–2003, 784), who cites the US's juvenile detention rate, murder rate, corporate crime, and trade deficit as evidence to contradict "the triumphalism of the [American] crusaders for Smithian economic liberalism." Todd and Fontaine have problematized the most widely accepted perception of the US – that of a powerful country.

Nonetheless, few other French authors are willing to second Todd's prediction that the US is close to collapse. Most settle for the safer, if less exciting, view that the US is strong but flawed, and perhaps a bit weaker than it appears. Calvez declares, for example: "Indications of the decline of the empire are not scarce: they refer to definite psychological and moral weaknesses; but nevertheless they do not appear capable of bringing about a rapid collapse" (2004, 23–4). Jean-Marie

Colombani traces more restricted signs of weakness in the US's need to "engage in individualized contractual relations with the emerging mid-level powers – China, Russia and India, and tomorrow no doubt with Iran and Indonesia" (Colombani 2002, 78). For François Duchêne, US decline remains in the future but will come as the result of unilateralism, as the costs in terms of lives and money will inevitably lead to pressure from within the US: "Multilateralism is not therefore a fetish of the weak. It is inscribed in the nature of contemporary problems including those which appear the most favorable to the exercise of traditional armed force" (Duchêne 2002–2003, 778). So while the US is not on the verge of collapse for the majority of French observers, it is facing the erosion of its power, even at its hour of global dominance, and will need to negotiate and compromise in order simply to survive over the long term.

Finally a worldview common in the US finds one champion in France. Jean-François Revel sees the US leading the global struggle for freedom. For Revel, as for American neoconservatives, the US is both powerful and good. It has the special vocation of promoting freedom by forcing open global markets. For Revel there is no difference between free market capitalism and political freedom; to critique one while enjoying the other is simply hypocritical and this minority opinion presents a counterpoint to the mainstream French view of the US. But the mainstream perspective, contrary to his allegations, is not so much "anti-American" as critical of selected aspects of American foreign and domestic policy and skeptical of American claims to moral superiority.

Conclusion

To take two influential critiques of the US as our guideposts, Védrine's view is the opposite of Todd's, yet both support responses to American power at the national and regional (European) level. The French as a whole are similarly divided in their admiration (mixed with dread) and their growing doubt (mixed with contempt) of the US. Very few wholehearted supporters like Revel can be found. What emerges most clearly from the debate is a picture of American global dominance, which even appears in a sort of ephemeral or illusory form in Todd's analysis. The disagreements primarily concern the implications of this dominance for France, Europe and the emerging global system. The most insidious threat is that US dominance may be forcing Europe's evolution on the route towards an unmanageably large and heterogeneous space devoted primarily to trade – a "space of flows" to adopt Castells' (1996) term – rather than a small and strong "space of places" defined by coherent political action and shared social values.

The building of the EU towards a federal model is viewed by proponents with a mixture of "Europhoria" and "Europessimism" (Borneman and Fowler 1997, 510). To a large degree this distinction is based on whether one sees Europe as an effective counterbalance to the US or as a victim of US manipulation. Europessimists in France fear that the poorer and newer members of the EU serve as Trojan horses for

US interests. Another sort of pessimism is the belief that in forming a union Europe has already given in to an excessively liberal social model (free market/weak state), conforming to American values and serving American interests.

While Védrine is right to argue that "opposition to the death penalty, the war in Iraq, or the Bush administration does not make one anti-American" (2004, 117) taking such positions does, in the current political context, help establish the US as an Other for Europeans. Consider the comment by Joseph Maïla: "From triumph to triumphalism, the step can be taken quickly, and above all by a country that has, in the name of ills that have befallen it, already violated the rules of shared decision making" (2003, 584). Within the "audience space" of the EU, reminders of this sort "manage" the problem of the pro-US stance of smaller and weaker members. So we might reply to Védrine that opposing US policies may not make one anti-American but does, in the French context, generally indicate that one is pro-European. More generally, calling for stronger international cooperation constitutes a stand against American foreign policy, which indicates the awkward position the US has put itself in by choosing the Hobbesian path of unilateralism.

The primary question at this point is the degree to which scholarly debates will infiltrate public, non-scholarly discourses in France over the next few years. To what extent will academic ruminations on a stronger Europe inspire "Europeanist" votes in France and a will on the part of the French public to share sovereignty with other European countries? How will *Europe puissance* and *hyperpuissance* motifs play out in the popular media? Will the US continue to appear as both strong and weak or will the picture resolve itself in one direction or the other? Inasmuch as the struggle against nationalism and for Euro-regionalism depends on the perception of an outside threat, the US will probably remain a key motif in the struggle between Europeanists and nationalists. Those intellectual observers ideologically inclined towards Europeanism will try to play up the US threat, because it adds to the apparent necessity of France as the leader of Europe. Nationalists will also use the US as a central symbol – in this case a symbol of the external forces that will control France if it continues to compromise with other European countries, specifically with small countries inclined to follow the US lead.

Many Europeanists who are deeply suspicious of US domination nonetheless see no point in trying to match American military capabilities while Atlanticists are content under what they see as a US umbrella. This bi-lateral bloc opposed to European militarization will likely ensure that the power Europe develops will primarily be economic and political, all the more so as remaining trade barriers are dropped within the EU and the EU increases trade with rapidly developing countries such as India and China. But something new is envisioned – a kind of soft ideological and cultural power exerted by Europe at the global scale, consisting of the ability to offer a model of social development that is democratic, tolerant, and humane.

French intellectuals almost universally support a move to rethink the polity on an expanded scale but they have not resolved the various questions about scale. Should France surrender more of its sovereignty to one, two, five, fourteen or twenty-four other states, or even more? Bigger does not necessarily mean stronger and if strength lies in coherence then the best union may be the smallest – a Franco-German republic. Between this minimalist pairing and the EU 25 lies a thicket of debate. US policy will doubtless influence this debate by increasing or decreasing the apparent legitimacy of pro-US governments in Europe and thereby affecting the strength of the guiding European nations – France in particular.

How important are these academic debates to the political life of a country? How large do they figure in France's civil society? Normally their influence is not great but occasionally the key terms and ideas of scholarly debates descend to earth and set foot in everyday discussions. For a fleeting moment, the motifs used in scholarly debates may be taken up by popular media to understand a particular situation or condition. Such moments are rare, and popular discourses never capture scholarly theories in their original depth. It would be a mistake, therefore, to view these scholarly writings as equivalent to "French opinion." Nonetheless, in autumn 2004, as the French anxiously watched the American presidential campaign, scholarly concepts like *Europe-puissance* and *hyperpuissance* circulated in informal communication contexts like Internet discussion forums and on television. The gatekeepers in the media and, even more, the French citizens, took certain cues from academic writing. The next three chapters provide excursions into a range of public discourses all of which contain echoes of these scholarly debates.

Chapter 5

Newspaper Reporting: Restraint and Balance

> A newspaper is an adviser that need not be sought out, but comes of its own accord and talks to you briefly every day about the commonweal without distracting you from your private affairs.
>
> de Tocqueville, *Democracy in America*, p.517

Newspapers occupy a special place among the panoply of news sources because of their relations to space and time, as well as their detail and their social role. As Gertjan Dijkink argues (1996, 143), "Domestic problems and events dominate the media's coverage of news and one has usually to turn to the few 'quality' newspapers to find adequate coverage of international events." Newspapers of national stature, like the *New York Times* or France's *Le Monde*, in particular, provide a long-distance newsgathering function and in-depth coverage of international events that is available only from a narrow range of information sources. The "raw material" obtained by foreign correspondents for major newspapers is then recycled in smaller newspapers, general circulation magazines, and television and radio news programs.

If newspapers can be said therefore to have a special relationship to news across space, they also have a special relation through time. They anchor the flow of time in an official version of the present, day after day, combining consistency with unusual detail. Newsprint deteriorates rapidly due to oxidation but this physical limitation is overcome by copying onto microfilm and computer files, or to be more precise, by the social institution of archiving newspapers in these ways. Newspapers thereby capture the character of the time, the zeitgeist, and historical events are remembered after many decades by their headlines. Even the most trivial of facts appear in newspapers as a matter of historical record, such as marriage notices and obituaries. The authority of the newspaper to define the present (and hence the past) derives in part, as well, from the diligence of newspaper editorial staff in cross-checking facts and striving for balance, factors which are less consistent in other media. It is precisely because of the exhaustiveness, care and thoroughness of print journalism that newspapers are treated as historical records, archived, and consulted by historians.

Poststructuralist, postmodernist, postcolonialist and feminist critiques have made abundantly clear the constructedness and the positionality of news stories (Darling-Wolf 1997; Berkowitz 1997). The rise in popularity of the Internet and its casually constructed, un-refereed discourses serve to put these critiques in perspective, however. In comparison to the Internet, newspapers excel in their detail and editorial attention. This does not mean newspapers are objective, but they differ greatly in their construction of reality from other media because of their divergent aims, objectives, and affordances. These combine with their geographical range to create special relations to time and space. Not all constructions of reality are equally rigorous, and the combination of in-depth coverage, details and geographical coverage affords newspapers more of a role in defining the nature of "objective" or "historical" reality than Internet Blogs and discussion forums (see Chapter 7), special-interest magazines, television news or radio talk shows. The latter media assume many other important meaning-creating functions, such as the forging of consensus or the structuring of debates, but newspapers offer one of the most solidly founded representations of reality beyond the compass of direct perception.

Newspapers also provide an indication of the attitudes and worldviews of readers: "A newspaper can only survive if it gives publicity to feelings or principles common to a large number of [people]. A newspaper therefore always represents an association whose members are its regular readers" (Tocqueville 1969, 520). Rather than objects they are social practices consisting of excluding and including, promoting and silencing, within the situated objectivity of a particular informal association or community of readers. Newspapers therefore take on an important norm-defining or "gatekeeping" role (Westley and MacLean, 1957) in society. In short, newspapers must be appreciated for their role in defining objective reality in both its spatial and temporal aspects, even if that reality is associated with a particular readership. Opinions will therefore be cloaked in the rhetoric of a report on what is rather than an attempt to persuade or incite action. Recognition of the difference between print journalism and the Internet, in particular, hinges on the crucial role of shared social constructions of history and geography within the entirety of social communication.

Newspapers are declining in importance as a result of the diffusion of the Internet, but they are unlikely ever to be fully replaced by other media because of their special social, spatial and temporal qualities. They are likely to evolve and come to occupy a particular niche in the complex communication environment of the 21st century. The French construction of the American election of 2004 through three newspapers reveals not just a vision of the US from France, but more specifically a serious and meticulous attempt to record a historical event in an unbiased way – to offer explanations and analyses rather than emotional impressions or opinions. While not completely free of bias, French journalism is less sensationalist, in general, than American, Canadian, or British print journalism, therefore the ideal of print journalism that I have presented is all the more valid in this study.

The chapter begins by introducing the three newspapers that will serve as sources; next it presents the major and minor rhetorical motifs in the newspaper coverage of the election; then it returns to consider the three newspapers in greater depth as well as some of the key differences between them.

The Sources

Le Monde, *Le Figaro*, and *Libération* are the three main daily newspapers with national circulation in France. In recent years all three have suffered declines in readership induced by the rise of television and the Internet combined with a failure to discover ways to build a dedicated readership among the younger generations. In France, just as in the US, newspaper readership has declined for several generations. The percentage of people who regularly read a newspaper in France is now lower than in 25 other countries, although France is the world's fourth largest economy. Bertrand (1999, 75) suggests several reasons for this situation, including the relatively high cost of French newspapers due to meager advertising revenues, the lack of delivery in certain regions, and public suspicions regarding the reliability of news reports. Yet another possible cause for the low circulation figures is that French journalism pays relatively little attention to the colorful "personal interest" stories that would cultivate a broad working and middle class audience. Finally, France's centralized government structure has produced conduits for obtaining information that are relatively pre-determined and journalists consequently often envision their role as one of interpreting information passed down to them from above. According to Bertrand:

> Seen from abroad, the French journalist appears to be a political commentator with literary ambitions. The stars of journalism, witnesses of history, high priests of information, consider themselves part of the intelligentsia. In the written press, they write for the elites, using terms and concepts unknown to the masses. They think (ancient myth) that they play a crucial role in political life.
>
> (Bertrand 1999, 78)

In content, style and organization, then, French newspapers lack broad appeal. The picture of French journalism is not complete without the inclusion of Agence France Presse (AFP), the only global news agency not based in the Anglophone world. The tendency of French newspapers to use this source means that their coverage of international affairs has a different emphasis than the news of the same events disseminated in the US, the UK, Australia, or English-speaking Canada. AFP coverage of the Israeli-Palestinian conflict, for example, regularly includes information and statistics related to Palestinian casualties while this information is sparse on other wire services.

Le Monde, with a circulation of about 400,000, is the dominant newspaper in the French market, and is taken by many as the best indication of "the French perception" of world events. Founded in the wake of World War II, *Le Monde* replaced the 83-year-old daily, *Le Temps*, decommissioned after the war due to links with the ousted Vichy government (Caslon Analytics 2005). Its political position is blandly neutral, its writing dry and prosaic, but both the absolute sales figures and the ten-year trend indicate a superior ability to maintain readership than is the case for France's other national circulation dailies. *Le Figaro* was founded in 1826 and today ranks close behind *Le Monde* in circulation, selling about 370,000 copies, although prior to World War II it was France's leading newspaper. *Le Figaro* targets an intellectual audience with conservative political leanings. Its stories contain nuances and complications that prevent its political leanings from approaching the glib propaganda that saturates popular American newspapers like *USA Today.* It boasts a surprisingly varied writing style ranging from the technical to the erudite to the poetic. So it is conservative but its political stance only partly defines its niche in the French media market. The third national daily, *Libération*, has been the voice of France's political left since its founding in 1973 by none other than the philosopher Jean-Paul Sartre. Despite France's powerful Left wing, its circulation lags behind the other papers, at around 170,000 copies, most likely because local dailies appeal more than a national paper to many on the Left. Its lively writing style (though not its politics) is vaguely reminiscent of American newspapers.

All three of these newspapers are, technically speaking, national papers, but sales outside of Paris account for only 27.5% of their combined sales total (Bertrand 1999, 75) because regional papers still dominate the markets outside of Paris. The coverage of international affairs in the regional papers is relatively meager, so interest in foreign politics is relatively low in rural areas of France, even regarding the hyperpower.

Ownership information consolidates our understanding of the political positions of these newspapers. *Le Monde* is produced by Groupe Le Monde, which also produces *Le Monde Diplomatique*, a monthly news magazine addressing international relations, conflict, and social issues, as well as a bimonthly extract from *Le Monde Diplomatique* called *Manière de Voir*. Groupe Le Monde has managed to maintain its economic independence, which is perceived by many in France as a necessary (though not sufficient) condition for unbiased and therefore high quality news. The prestige held by Groupe Le Monde's publications has not prevented "accumulated losses of several million dollars" (Caslon Analytics, 2005) an unthinkable situation for the intensely profit-oriented American newspapers. *Le Figaro* does not have such worries. France's aerospace giant, Groupe Industriel Marcel Dassault owns a controlling share of the conservative daily. Dassault bills itself as "Europe's leading exporter of combat aircraft" and sells its military aircraft to India, Peru, Greece, and various countries throughout the Middle East (Dassault 2005). Corporate control is not the same as censorship, but it clearly affects editorial policies. *Libération* is under more distributed control, with considerable economic autonomy. "As of August 2004 around 21% is held by [the film and television giant] Pathé (down from 64% in the late 1990s), 23% by staff associations (primarily its leading journalists) and 14% by a holding company whose members include 'celebrities'" (Caslon Analytics, 2005). This ownership pattern again conforms to the newspaper's politics as it promotes autonomy from government and (non-media) corporate influence.

The three news sources differ, therefore, in terms of their political positions, style and relative autonomy from external control. Generally differences in news coverage reflect differences in the social position and the organizational autonomy of newspapers, but on any given issue, such as the US presidential election, it may be difficult to ascertain a clear cause and effect relationship between the motifs in a particular article and a newspaper's funding and affiliation. The ramifications of a foreign story like the US election may not be so clear as to generate a single coherent journalistic response in a particular newspaper. The differences between these sources merit further discussion but this will be delayed until the final section of this chapter in order to briefly introduce the motifs that structured reporting in all three newspapers.

Motifs

Two themes dominated the newspaper coverage of the 2004 election: (a) the unusually deep rift between opposing factions of the American electorate and (b) the pivotal role of violent conflict (September 11, terrorism, Iraq) in shaping the mood of the American electorate. These major themes are considered below as rhetorical motifs, under the headings *Divided America* and *Nation in Shock* (motifs 1 and 2) and situated among four other related motifs. It was clear in all three newspapers that barely half of the American electorate supported Bush and that his victory took place in the context of a profound shock that precipitated uncertainty and fear, sentiments which could be driving a wave of nationalism. Yet if the US was responding to a severe shock, it was also evident that not every nation would respond in the same way to such a shock, and that the religious overtones to this response were a sign of American peculiarity, particularly America's *Fundamentalist Society* (motif 3). Fundamentalism was all the more intriguing in light of the materialism of American culture, which in turn had driven the economy to its position of global dominance because the French associate materialism with secularism. Yet this dominance was looking precarious owing to the accumulation of debt, suggesting the specter of a *Collapsing Economy* (motif 4). The campaign coverage naturally included a motif we could call *The Candidates* (motif 5) but this motif was deeply ambivalent. The bitter divide in the American electorate did not necessarily mean that one candidate would be better than the other in a practical sense, for French workers, citizens, companies or politicians. Kerry looked more *sympathique* (understanding and understandable) but that might not be a good thing for France, since antagonism with the Bush administration was, in general, a useful state of affairs for French politicians (the latter being primarily Europeanists and benefiting from a breakdown in transatlantic relations, see chapter 4). Therefore the final motif is *Useful Threat* (motif 6) meaning the US political mood viewed from France through the lens of local political objectives and ambitions as both irresponsible *and* useful. These motifs are summarized in Table 5.1 and associated with various minor motifs.

What emerged from the French newspapers was far from a scathing critique of American society. Instead, journalistic discourses worked to normalize relations between the countries and to "claw back" representations of the US to well-established frameworks that were not always flattering to the US but were far from anti-American. Only *Fundamentalist Society* verged on anti-Americanism. Surprisingly, the modes

Table 5.1 Major and minor themes of the French newspaper coverage of the 2004 American presidential election.

Major Motif	Minor Motifs
Divided America	extraordinary political divide wide range of disagreements authoritarian Republicans and ridiculous Democrats
Nation in Shock	politics driven by fear both parties seeking "war president"
Fundamentalist Society	Catholic/Protestant divides breaking down Democrats more secular
Collapsing Economy	poverty in the US wealth concentration debt
The Candidates	no difference between the candidates big difference between the candidates
A Useful Threat	US intransigence useful across the political spectrum US intransigence useful to nationalists and Europeanists

of journalistic representation generally worked *against* the deterioration of Franco-American relations that had been taking place over the previous four years.

While consideration of the differences between the papers will be the focus later in the chapter, it is important to note up front that the approach of *Libération* was somewhat distinct from that of the other two newspapers as it grew from a fundamental sense of common cause with the American left, and a growing sympathy in particular for the Democrats and for John Kerry. Like the other newspapers, *Libération* limited anti-American sentiments by linking America's apparent failures, excesses, and weaknesses with a segment of the American population rather than with the population as a whole. It went a step further in quickly taking sides. The newspaper openly supported John Kerry despite the overt admission at several points that a victory by Kerry would almost certainly pose greater strategic problems for the left in France and in Europe than would a victory by Bush (*Useful Threat* motif). The other two newspapers were slow to reveal an editorial preference and remarkably circumspect in the ways they did so. The massive preference of the French public for Kerry was scarcely evident in *Le Monde* or *Le Figaro*. None of the three papers carried this revelation a step further to the point of anti-Americanism. In general, favoritism for Kerry was revealed by critiquing Kerry on matters of tactics and strategy while critiquing Bush on matters of judgment and honesty. For example, in an article entitled "Iraq undermines the party," *Libération*'s American correspondents argued that the Iraq issue must be handled very cautiously by the Democrats because of their internal divisions over the war, Kerry's inconsistent voting record, and the fact that withdrawal from Iraq would be a slow process in any case. An editorial in the same edition of *Libération* suggested that Kerry should adopt the following strategy:

> To win this challenge, Kerry must do more than get his shirt sleeves wet. He must get himself entirely soaked, because thus far he has barely profited from the setbacks of his rival in the White House, who is tied with him in the opinion polls. If this uncertainty makes the election more interesting, it is disturbing for Kerry. He therefore has to land on Bush not only on the economic terrain and on interior security but also with regard to Iraq, where the left faction of his party (whoever is close to Ralph Nader or Michael Moore) reproaches him for wavering and lacking political courage"
>
> (Gaudemar 2004b, 3).

Kerry came in for criticism, of course, but the focus on critiquing campaign strategy indicates clearly that he was the favorite.

Divided America

As early as July 30, French journalists were mentioning "The electoral campaign… that could well be one of the most polarized that America has known in recent years" (Jaulmes 2004b, 2). The sense of a deeply ruptured America resonated throughout all three of the French newspapers during the build-up to the election. *Le Monde*'s Serge Marti described: "an America organized into two camps separated by seemingly everything – the war in Iraq, the 'values' and heritage of the founding fathers, grand social and economic choices – but also the type of relations that the world's primary political, military, economic and cultural power should maintain with the rest of the world" (Marti 2004). Patrick Sabatier of *Libération* attested that: "Two Americas clash, divided and more polarized than ever in regard to geographical, ideological and moral, economic and social viewpoints, and in their relation to the rest of the world" (Sabatier 2004). *Le Figaro* sounded a similar note: "in regard to the course of the antiterrorist war, the management of international alliances, the orientation of society on subjects like abortion, economic policy, energy, the environment, and scientific research… On all of these questions, the radical disagreement between Republicans and Democrats fuels a passionate campaign" (Gélie 2004b, 2). An article in *Libération* entitled *Deux Amériques dans un mouchoir de poche* which is to say, two Americas in a pocket handkerchief (the French expression for a very close race) (Riché 2004c), captured both the motif of social division and the idea of a dead heat in the election. An article on John Edwards in *Le Figaro Magazine* proclaimed that: "Whatever may be the outcome of the presidential election of November 2nd, the campaign that has concluded will have marked the political landscape of the United States as one of the most disputed of its history" (Debat 2004, 20). Countless stories about various aspects of the campaign drew on this motif of profound division, such as the September 30 story from *Le Figaro* called "The Savage Ad War" (*La guerre féroce des spots publicitaires*) (Faure 2004b).

The French press took its cues from politicians on both sides of the Atlantic. Jack Lang, the Socialist representative from Pas de Calais spoke of: "[T]wo American cultures, two visions of society, two conceptions of humanity in collision" (quoted in Sequence France, 2 Nov, 2004: 7). John Dean, former counsel under President Nixon administration, confirmed this schismatic perception in a half a page op-ed piece in *Le Figaro*: "After forty years dedicated to the study of American presidencies, I must note that I have never seen the United States in the grip of such a political division" (Dean 2004, 12).

The motif of division, confirmed by American and French observers, gained momentum as it circulated in the French press and worked against any reduction of America culture to a single set of attitudes and beliefs, *sypathique* or alien. Likewise it blocked the emergence of anti-American representations. Although the motif of social division was not, in itself, necessarily reassuring, French observers outraged, frightened, disgusted or annoyed by George Bush might find solace in the fact that close to half of the American population shared their feelings. Perhaps the motif of fragmentation also weakened America in the eyes of foreign observers, reassuring those who feared the hyperpower that it was not a unified power (Fig. 5.1). But the motif inspired a closer look, as we shall see, at the people and places associated with the two sides.

While the attitudes and electoral posturing of the Republicans closely matched the stereotypes that have long circulated in France's anti-American discourses, they provided at least the satisfaction of familiarity. The image of swaggering cowboy-cum-businessman that Bush cultivated topped any stereotypical American the French could have produced. This in turn encouraged a reassuring sense of cultural superiority. Attitudes of the Democrats, in contrast, clashed with anti-American biases and made the US appear to be a more ambiguous place.

If the US was in the midst of a major political struggle that produced a sense of uncertainty about its course and therefore an ambivalent image of the global dynamics in which France would need to navigate, the Republicans were a known quantity because they were quintessentially American in French eyes – inclined to favor business, hostile to the arts and intellectual pursuits, materialistic, hawkish, pro-gun, and overtly demonstrative of their religious views. The Democrats failed to fit the mold of American culture as viewed from France or the norms of French culture. The Democrats in fact struck the French as rather ridiculous, for example when they struggled to capture the anti-Bush passion ignited by Michael Moore's *Fahrenheit 9–11* while at the same time concealing his presence at their convention (Faure 2004a, 3), or when they interpreted the Red Sox victory as a prophecy of Democratic victory (Jaulmes 2004a). These strange signals from the American Left provoked bemusement as often as sympathy, particularly in *Le Monde* and *Le Figaro* where they were presented as bizarre and ridiculous. In contrast, *Libération* expressed unalloyed enthusiasm for the American Left and chose to see their struggle as its own.

Figure 5.1 America divided cartoon
Source: *Le Figaro* 3 November, 2004, p. 6, with permission.

The "undecided voter" was at once the exception to the rule of *Divided America*, and yet also the focal point of that division. Which way would these oddly neutral Americans be more inclined to tilt when the time came to vote? Just as this question weighed heavily on the minds of Americans it also weighed on the minds of the French (fig. 5.2). The schism in the American electorate struck some observers as the result of outright manipulation. Patrick Jarreau of *Le Monde* blamed the ideological divide on Bush's uncompromising political approach:

> The Americans are not unified in response to exterior dangers or interior questions. They are divided on the one as much as the other. In place of responding to the demand for

Figure 5.2 An American voter holding the world in his hand; the label on his shirt reads "undecided." Denis Pessin, *Le Monde* 2 November, 2004, p. 1.

> security in a way that permits him to expand his support in the country, Mr. Bush has turned it into a partisan question. This strategy may bring him the votes he needs in order to ensure his reelection. It might also cause him to lose.
>
> (Jarreau 2004c, 15)

In short, Bush appeared to be the cause of the rupture in the American electorate. The same note was sounded by Adrien Jaulmes (2004b, 2): "the personality of incumbent president George Bush is detested or adored in equal proportions by popular opinion; the collective reflexes of the United States post September 11 remain those of a country at war." Here we see the intersection of the *nation divided* motif with another motif of great importance: the motif of *Nation in Shock.* For many French observers, the polarization of the US was best understood in light of the galvanizing effect of the September 11 attacks.

Nation in shock

French coverage of the election was framed with great consistency in terms of the idea that the election was an expression of the fears and passions ignited three years earlier by the September 11 terrorist attacks. Headlines like "The September 11 attacks are at the heart of the American presidential campaign" (Jarreau 2004a, 3), "Terrorism at the heart of the electoral duel" (Gélie 2004f, 2), and "Bush takes up the standard of 9–11: He wants to establish himself as the only leader incarnating 'the courage of the nation'" (Riché and Rousselot 2004) emphasized this motif. The idea held special significance in France because it signaled that whatever induced Americans to take extreme positions in international relations might be only a temporary condition, and relations between the two countries stood a chance of normalizing. If the majority of the American electorate chose a candidate generally despised in France, their choice might indicate a reaction to profound and almost unimaginable shock. This psycho-social diagnosis certainly reduced in some measure the tendency of the French to judge Americans negatively for supporting Bush.

In an interview printed in Le Monde, François Holland, First Secretary of France's Socialist Party, observed America's situation with sympathy: "First off, try to understand what has happened. Since September 11, America is a country at war. Its army is engaged in Iraq and suffers important losses every day. In these circumstances, the temptation to join forces behind the incumbent president is particularly strong, despite his lies, errors or oversights. The increase in [voter] participation therefore benefited above all the Republican candidate and the reflex of fear" (Holland 2004, 12). He continued, however, by shifting from sympathy to realpolitik and observing that Bush's victory provided an opportunity for strengthening Europe, since other European countries would perceive the futility of negotiating with an America led by Bush and would instead turn to each other.

The Democrats were not immune to post-traumatic shock. It appeared that Kerry was chosen by the Democrats as the "most serious alternative to Bush in the role of 'war president' (Jaulmes 2004b, 2). The Democrats apparently could not succeed if they did not stress American toughness. "The pacifist stances of Howard Dean, much more popular among the most liberal sectors of the Democratic party, run the risk of translating into massive rejection on the part of the electorate" (Jaulmes 2004b, 2). The mood of the post 9–11 electorate presented a special sort of challenge for the Democratic candidate and the French newspapers were clear on this point. They stood on the sidelines alternately cheering him on, critiquing his tactics, and offering advice. All of this attention illuminated his challenge, which ultimately helped readers understand his defeat. Rather than confirming suspicions about the US as a new empire likely to strike out in any direction at any time, this motif suggested a passing mental affliction.

The *Nation in Shock* motif was not invented in France. French media took their cues from American media on this narrative, but it figured largely in French newspaper reporting because of the role of newspapers in maintaining familiar worldviews and "clawing back" the unexpected and inflammatory events to familiar worldviews (Fiske and Hartley 1978). America has many faults in the common French view, but launching unprovoked attacks is not one of them (thus the surprise in 2003). *Le Monde* featured an interview with Corey Robin, a political scientist from City University of New York, who believed that the Bush administration was governing and campaigning primarily through the provocation of fear (Robin 2004, 4). Other interviews and editorials of American scholars and statespersons confirmed this view. Clearly the French perceived security issues differently from Americans – whether they were thinking primarily of America's security, their own security, or global security – but highlighting the US's security concerns could help situate Bush administration policies as symptoms of recent events rather than deep problems in American society. This tendency to see the election as a sign of troubled times rather than a measure of America's soul or essence was not universal, however, as indicated by the sub-headline on a November 5 article in *Le Monde* (Leser 2004b, 5), according to which: "The election indicates that the United States is more conservative than observers had thought." But most stories attributed the conservative swing to extraordinary conditions rather than the fundamental character of the US.

The *Nation in Shock* motif also helped distinguish between the two candidates. *Le Monde*'s Washington correspondent contrasted the Bush administration's "war on terrorism" with Kerry's call for better policing, intelligence gathering, diplomacy and ideological engagement (Jarreau 2004a, 3). Kerry's proposal met with cautious approval from the French newspapers while Bush's approach was viewed as a sign of what François Holland called a "reflex of fear" (Holland 2004, 12). Distaste for Bush's militarism was aggravated by the incumbent's self-proclaimed "born-again" Christian faith. If America's response to shock was somewhat comprehensible, the religious element of America's response to the shock was not.

Fundamentalist society

American news coverage of the 2004 election placed the issue of "values" above national security. One would expect "values" to figure into the French coverage of the election insofar as the French news media were taking many cues from the American media. The French media could be expected, as well, to regard these "values" critically since they were in fact religious beliefs and Americans have been seen by the French as religious extremists for over two hundred years (see chapter 3).

As early as August, the force of religious traditionalism as a dimension of the campaign was recognized in France. A headline in *Le Monde* read: "Christian extremists campaigning for Bush" and quoted Alfred Ross, of New York's Institute for Democracy Studies, who argued that "the networks of the religious right are at the very heart of Republican power" (*Le Monde* 2004a). Sensitivity to the religious dimension of American politics increased throughout the fall. In October it was linked to prospective nominations of Supreme Court justices (Gélie 2004e), to the erosion of support for Republicans among gay voters (Vincent-Arriola 2004), and to Kerry's loss of support among many Catholics for defending the right to abortion

(Faure 2004d). All of these factors were presented as peculiarities of the American political scene. Photographs of Bush and Kerry appeared side by side in *Le Figaro*, each apparently in church and each with his hands raised in a gesture of supplication. The caption for these images (*Le Figaro*, 29 October, 2004: 4) read: "For George Bush (left) and John Kerry, both Catholics, going to church is a political obligation: piety plays an important role in the US in the conquest of diverse communities of voters." This mistake regarding Bush's denomination was not repeated elsewhere, and the ability of Bush to win over conservative Catholics from his rival was described elsewhere as an indication of successful "political marketing" (Gélie 2004g, 4). Here is a striking continuity with much older French visions of the US; as Philippe Roger argues, Americans have long been too religiously innovative to please the French clergy and too religiously devout for the taste of the rest of French society. Furthermore, Americans have been viewed as applying the principles of commerce indiscriminately to all spheres of life. So it was natural for French writers to speculate that Bush's "capture" of Catholic voters was brought about by a self-conscious attempt to market himself to America's *fous de dieu* (religious fanatics).

In a post-election assessment, Patrick Jarreau of *Le Monde* summed up Bush's "conservative revolution" as dryly and dispassionately as possible, simply quoting Bush's plan to "maintain the deepest values of the family and faith" (Jarreau 2004f, 2), but other authors were less restrained on the religious issue. Henri Tincq, also of *Le Monde*, saw evangelical Protestants and conservative Catholics as united in their "messianic" vision of America. "This social and moral conservatism, a mix of patriotism and religious fervor insisting on the messianic role of America in the global battle against evil, has become their principal electoral reservoir. The evangelical electorate sees itself in this Manichean vision, in the mission it has set for itself to shape politics according to 'a biblical vision of the world,' as Tom DeLay… says" (Tincq 2004, 6). He continued by pointing out the irony that Bush, a Protestant, had become the Pope's main ally in the fight against abortion, stem cell research, and gay marriage. "On these themes, pro-Bush Catholics orchestrated a violent anti-Democrat campaign. Ultraconservative bishops maintained that voting for Kerry would be a sin, forgetting that John-Paul II was one of the fiercest opponents of the war in Iraq" (Tincq 2004, 6). Democrats were described as "dumbfounded" by the realization that 61 percent of voters who attended church at least once a week voted for the President (Lesnes 2004b). Like Tincq, Corinne Lesnes of *Le Monde* had difficulty reconciling the idea that hawkishness and religious fervor are strongly linked in the American political terrain. The essayist Guy Sorman published his interpretation of the election in *Le Figaro*, arguing that fundamentalist Christianity placed solidarity with Israel beyond debate: "less by reason of the 'Jewish lobby' as written elsewhere than as a consequence of the Americans' identification with the Hebrews of the Old Testament and their rather widespread conviction that the US is, like Israel, a promised land, and the Americans, a chosen people" (2004, 12).

The increasing power of the Republican Party and its ties to fundamentalist Christian movements provoked a kind of anxious fascination in France. More than any other aspect of the Republican Party agenda, including militarism, the religious dimension was treated unsympathetically by the French Press. The front page of *Le Monde*'s August 29–30 issue bore the headline "Christian Extremists Campaign for Bush" and the accompanying cartoon showed Bush telling his advisors "I propose that we start the day with a little prayer for my reelection." This frankness was perhaps accentuated by vagueness in other areas: the candidates did not give clearly differentiated signals on US military involvement, economic differences were even trickier to define, and the French response to these economic issues was deeply ambivalent. Whereas the French Left might sympathize with the US Left, French workers might actually suffer more in the case of a Democratic win. In contrast to these murky issues, the religiosity of the Republican rhetoric was easy to identify and easy to critique in the French framework of political debate. To better understand the ambiguity of the economic motifs mobilized by French media in connection with the election we turn now to the motif of "collapsing economy."

Collapsing economy

The American economy was a source of universal critique. From the Left (i.e. *Libération*), the problem appeared in the form of large and growing sectors of the American population living below poverty line, without insurance, and with ever-diminishing opportunities. From the right and center (*Le Figaro* and *Le Monde*) the US economy as a whole appeared to be teetering on the brink of collapse due to the combination of the international trade deficit and the federal budget deficit, which presented a threat not only to the US but also to the global economy. Yet one senses in these gloomy forecasts an unspoken interest in American economic collapse because of the opportunity it would present Europe to regain the dominant position in the global economic system that it had held prior to the 20th century. French business interests might revel in US economic weakness if it meant an opportunity for Europe to more successfully compete in the global marketplace, but US economic collapse would almost certainly have negative repercussions in Europe, as well. Thus pity, opportunistic interest and risk aversion could all be discerned in this economic motif.

Conventional wisdom on both sides of the Atlantic held that the American economy should be a decisive issue in the 2004 election. "Close to 36 million Americans live beneath the poverty line," noted a headline in *Le Monde* on the 28th of August (Stern 2004, 3). The article specified that the poverty rate in the US had increased since 2000, that the number of people lacking health insurance had increased to 45 million, and that in the past year child poverty had risen to 17.6 percent of children: a "spectacular increase." The article also provided quotes from Kerry accusing Republican policies for the bad economic situation and a quote from a Republican senator arguing that Kerry's plan would only make things worse. The article forecast that the economy would play a major role in the election but discretely avoided predicting which party would benefit.

A few days later, Philippe Gélie of *Le Figaro* was less discrete. Normally restrained and detached, he declared sardonically: "George W. Bush's social program would be centered around an 'ownership society' in which everyone would be encouraged to own their own housing and also to command their own retirement and health care funds through a partial 'privatization' of social security. In a country where 45 million people have no health insurance and where 36 million live below poverty line, you have to be a dedicated Republican to believe this" (Gélie 2004a, 5). *Libération*'s Washington correspondent reflected that "The economic situation not being fantastic, John Kerry could have built his campaign on only this theme as Bill Clinton did, with success, in 1992 ('the economy stupid' he repeated). Kerry did not make this choice" (Riché 2004d). From across the political spectrum the French assumed the economy was the top concern for the average American and its problems were legion: diversion of wealth to the wealthy, massive federal budget deficit, chronic unemployment, stagnant salaries, escalating health costs, growing ranks of uninsured, and dependency on foreign fossil fuels. Similarly, the main article of *Le Monde*'s November 3 "Economy" section was dedicated to the economic changes during the 2000–2004 period.

> The most fortunate can only congratulate themselves for the colossal tax cuts that have permitted the richest Americans to receive, in four years, incomes eleven times higher than the poorest of their compatriots. The ranks of the latter, in contrast, have swollen. They now represent 12.5% of the working population and the country has close to 8 million working poor...Justifiably it is on employment and job losses that the candidate Bush is least comfortable. Therefore, as his opponents hammer in, he is the first incumbent president in sixty-one years to show a negative balance of net job growth, in the form of 821,000 jobs lost during his term in office.
>
> (Marti 2004a, I)

Across the political spectrum the US's economic situation looked grim. The same article continued with the observation that John Kerry wanted to take control of the US's economic dysfunctions while Bush planned to pursue further tax cuts even if that meant adding to the cascading deficits – budgetary, trade, interior and exterior – that were weakening the American economy. Despite this difference, *Le Monde* suggested that the implications for Europe might be the same: either candidate would likely implement protectionist trade policies as a response to the trade deficit and loss of jobs (Leser 2004a). Yet from *Libération* came a different interpretation. "The positions of the candidates on the economy are more divided than on foreign politics. Kerry considers that the government has a key role to play when the economy passes through a difficult period, and has the responsibility to put in place a decent social safety net. Bush, fierce partisan of the market economy, proposes a radical project, that of transforming America into an 'ownership society' in which the government would have a minimal role" (Riché 2004d). *Le Figaro*, with its significant connections to big business, took a more analytical approach to the economic question, revealing the strong Republican bias in the poorest county in the US (Faure 2004c) but also predicting a more significant role for economic concerns in the swing states than would eventually be indicated by the election results (Gélie 2004d).

The two presidential candidates figured in the economic motif, as they figured in the other motifs. But in turning at last to what would seem to be the focus of the news coverage, even the master motif of the election, *The Candidates*, we find a kind of empty center. This motif lacks a clear perspective except in the case of *Libération*, which took an unambiguous stand on its preferred candidate.

The candidates

In the French press, Bush and Kerry were sometimes contrasted, sometimes portrayed as virtually identical. The two representational motifs employed to represent the candidates can be captured in the terms *No Difference* and *Big Difference*. When viewed as representatives of the Democrats and Republicans, through the motif of *Divided America*, the two candidates represented two visions of America. These visions, in turn, implied opposite policies at least on the domestic front. However, when viewed from a more calculating and pragmatic perspective of realpolitik and global trade, the differences dwindled.

An article by Claire Tréan in *Le Monde* revealed both motifs. She started the article by noting that "Never has the outcome of an American election been awaited with such impatience by the world and never have we hoped so much that the voters will remove the standing president from the White House. In regard to unpopularity outside of American borders, George Bush has attained scores unequaled in the history of the United States" (Tréan 2004b, 8). Later in the same article, however, she noted that a retreat from Iraq did not figure in the plans of John Kerry any more than in the plans of George Bush, despite France's calls for such a retreat. She noted, furthermore, that "In other international concerns and in commercial negotiations, John Kerry would be no less underhanded than, for example, Bill Clinton," adding that the only thing to expect from Kerry was a return to pragmatism in American foreign policy (Tréan 2004b, 8). Thus, she starts with *Big Difference* and ends very close to *No Difference*.

There was little disagreement about the incumbent among the French newspapers. Bush's "Texan" image, perceived not too differently in its basic elements whether one reads French or American newspapers – macho aggression, rhetorical clumsiness, disdain for the arts and sciences, demonstrative piety – was inflected in France by rather different connotations of these same traits. Most French writers constructed these as weaknesses rather than positive qualities. More importantly, as a force in geopolitics these "Texan" qualities were believed to constitute a threat to global stability. In contrast, there was considerable ambivalence about the prospect of a John Kerry presidency. The newspapers all mentioned his inconsistent voting record and ambiguous positions as well as his failure to produce significant legislation. In doing so they drew almost entirely on portrayals of the Democratic candidate that were circulating in the American media, shaped by conservative corporate interests. These factors rendered Kerry an unknown quantity in France, although all three of the newspapers eventually demonstrated pro-Kerry sentiments, with *Libération* earliest and clearest in this regard.

A particular source of interest for the French press was the "European" quality of Kerry's image, produced by his family ties, international experience and ability to speak French. Kerry failed to fit the negative stereotype of an American leader as semi-literate, uncultured, and trigger-happy. He thereby disrupted conventional geopolitical representational motifs, prompting a certain amount of interest, but somewhat more suspicion. Bush was a known quantity; he fit the American stereotypes and was comforting for that reason. Behind Kerry's vaguely European exterior there *must* lurk a hidden, and therefore dangerous, American interior, one all the more dangerous for being hidden. Displeasure with an alien Other enhances one's sense of national identity; so deviations of that Other from the execrable stereotype are disturbing to a sense of self and nation.

Compounding this symbolic discomfort rooted in the French perception of Democrats in general was a practical concern: if Kerry would be more diplomatic in his dealings with Europe he might eventually succeed in rebuilding Atlantic alliances between European countries and the US. That, in turn, would undermine efforts within Europe to forge a stronger European Union. Kerry's multilateral and cooperative rhetoric might resuscitate the spirit of Atlanticism at the expense of European consolidation and autonomy. Oddly, the editors of *Libération*, who represented the left wing parties with the most to lose in the case of renewed Atlantic cooperation, were the most favorable towards the Democrats. Presumably their ideological leanings overcame their strategic calculations. They simply judged Kerry as a leader, vicariously taking the position of American citizens, rather than calculating his potential impact on their own economic interests.

Coverage of the two candidates, like the rest of the election coverage, was enlivened by the peculiarity of American culture and politics. An editorialist from *Le Monde* volunteered, "My impression of the two candidates, purely psychological, is that one, George Bush, was limited but self-assured, and that the other, John Kerry, appeared more capable, which is sometimes a political handicap" (Casanova 2004, 6). A country where ability is a handicap must surely be a strange place. It is strange to the French, as well, to contemplate a place where a body-builder and action film star like Arnold Schwartzenegger can be used by a political party to generate an aura of "compassion" (Faure and Gélie 2004, 5).

A few articles, mainly in *Libération*, expressed unalloyed enthusiasm for a Kerry presidency, but most French newspaper articles demonstrated their preference for Kerry by critiquing him. Favoritism was evident in that the critiques proceeded in a somewhat different manner than the critiques of Bush; Kerry was criticized on tactical grounds for not fighting hard enough or well enough, for failing to clarify his positions or for inadequately pushing his advantage and Bush was criticized, in contrast, on the basis of his attitudes, beliefs, and policies. The pertinent distinction was between strategic errors on the one hand and dangerous stupidity or cupidity on the other. The *Big Difference* motif might therefore be expressed through open approval for Kerry, or more subtly through the *kind* of critique aimed at Kerry, that differed from critique directed to Bush. Enthusiastically pro-Kerry views contended quietly with guardedly pro-Kerry views, and both of these contended with views rejecting both candidates as unsatisfactory.

We can add to this picture by noting several details. First, *Big Difference* was particularly prevalent in *Libération*, where the framing of Bush differed quite noticeably from that of Kerry, while *No Difference* was common in *Le Monde* and in *Le Figaro*, but was mixed unpredictably with *Big Difference*, resulting in an incoherent portrayal of Kerry and, more generally, of the Democratic party. Pascal Riché of *Libération* wrote: "Whether on the war in Iraq, the role of the United States in the world, budget policy, or major subjects of society (health, abortion, medical research, the place of religion, rights of homosexuals…) the two candidates have strongly opposing positions" (Riché 2004c). Many articles in *Le Monde* and *Le Figaro* noted the divide in the American electorate but stopped short of proclaiming the Democratic candidate as the true representative of America's political left. Sometimes the basis of differentiation was erroneous, like the claim in *Le Figaro* that Bush's childhood family was "infinitely more rich and powerful" than Kerry's childhood family (Gélie 2004c, 3).

Second, the tentative support of Kerry solidified during the last two months of the campaign as French media were able to sort his viewpoints out of the patriotic posturing. This journalistic drift toward the Kerry camp also drew energy from survey findings (*Figaro* 28 October, 2004; Tréan 2004a, 18) that showed strong support for Kerry in France. What is particularly interesting in this regard is that the pro-Kerry stance of the French newspapers followed the shift in public opinion rather than leading it.

Third, disapproval of Bush was expressed in newspapers with rhetorical latitude, a freedom of expression that indicated a viewpoint so taken-for-granted that delicacy was no longer necessary. For example, *Le Figaro*'s editorialist Renaud Girard referred to Bush as "autistic" (2004, 13), which is strong language, and it must be noted that he was echoing the French President Jacques Chirac (Fulda 2004, 3). In Kerry's defense, *Le Figaro*'s Philippe Gélie (2004c, 3) commented that "every nuance is ridiculed by the Republican camp as a sign of weakness" and continued, "John Kerry preaches – sometimes in the desert – that the greatest power in the world would be stronger if it reconnected with its allies." And Gélie's exasperation is clear when he mentions Bush's ability to personify "the little-guy, the independent entrepreneur, the cowboy of simple tastes" (2004c, 3). Yet despite the difference in image, there was a lingering unease that crossed the party line, "What is striking, to start with, is the consensus of the two adversaries: the United States [is the] 'world leader' and the question is, at best, how to improve this sagacious authority" (Cohen 2004). Yet for this translator and writer, published in *Libération*, "The spectacle of these two white Christian men, one relatively intelligent, the other downright simian (*carrément simiesque*), is all the more gripping because they claim not only to 'lead' a world about which they know nothing at all, but also America, of which they are completely unrepresentative" (Cohen 2004).

The blending of *No Difference* and *Big Difference* motifs had much to do with the fact that French opinion leaders did not choose sides in a way that was easily understood. If we want to come up with a simple prediction of their response all we have to do is note that the French political spectrum is wider than that of the US and shifted substantially to the left. "Middle of the road" in France is therefore equivalent to the Democratic position in the US while American "middle of the road" corresponds to the position of the French Right. Because of this difference between French and US political positions, the French socialists who dominate the left might well dismiss the Democrats as scarcely any different than Republicans since both are positioned to their right, while centrists in France might be expected to welcome the Democrat with open arms. This was not the case. A *Le Monde* article of November 2 (Séquence France 2004, 7) characterized the attitude across France's political spectrum as "anyone but Bush." The subtitle of this article explained ambiguously: "If they had to declare a position in the American presidential election, Tuesday the 2nd of November, French politicians on all sides would vote mainly for John Kerry. Many of them nevertheless deplore the similarities between the Democratic candidate and the incumbent." *No Difference* and *Big Difference* are again clearly contending here. The principal support for the *No Difference* position comes from Maurice Leroy of the centrist UDF party who proclaimed that "the difference between Kerry and Bush is the thickness of cigarette paper." Alain Krivine of the Trotskyite LCR party concurred that "the drama is that between these two there is no true choice for society," and a Green Party representative, Noël Mamère, admitted reluctantly that he would feel "obligated to vote for Kerry even if dragging my feet." Meanwhile, several socialist politicians who were interviewed for the story expressed a strong preference for Kerry: "In Paris people often turn up their nose at Kerry," lamented Jack Lang, adding: "I would vote against Bush, but also for Kerry, and without a second of hesitation." His support is echoed by another socialist, Jean-Christophe Jambadélis, who calls himself "One hundred percent pro-Kerry." Meanwhile Jean-Marie Le Pen, the neo-fascist candidate who electrified France's 2002 election and, by frightening the Left, ultimately opened the door for Jacques Chirac (from the right wing UMP party), stood almost alone in his preference for Bush. In short, Kerry's strongest supporters were not those with positions closest to his, but rather those of the French left who were well left of Kerry himself. Meanwhile, the French center and the Greens strove to maintain neutrality despite the similarity of their political stance to that of the Democrat by promoting the "no difference" motif while the far Right sided predictably with Bush. Newspapers looking to these political leaders for ideological guidance would find a contradictory set of guideposts.

Comprehension of Kerry's position improved throughout the duration of the campaign. Taking a sample from mid-summer, an article in *Le Figaro* argued that "At 14 weeks from the presidential election, the image of the Democratic candidate is all the more scrambled because his political rhetoric has remained as fuzzy as possible since the beginning of the campaign" (Jaulmes 2004b, 2). Continuing in the same critical tone, the article complained that, "In its general outlines, John Kerry's electoral program might easily be summarized as: 'almost like Bush, but better, and above all without him'" (Jaulmes 2004b, 2). The snide tone reveals an underlying sense of disappointment with Kerry for failing to provide a clear alternative to Bush. Early in the race, Kerry was criticized, among other things, for expressing the will to use force "if necessary" in international relations and for stating that he would not give another nation or organization a veto right over US national security. Even *Libération*, which would eventually abandon all pretense of neutrality with regard to the American election, published comments critical of Kerry in July. An editorial by Antoine Gaudemar (2004a) expressed disappointment in the failure of Kerry and Edwards to articulate strong positions: "Their way of presenting themselves thus far as the champions of the middle class and of the values of middle America hardly clarifies things" (Gaudemar 2004a, 3). An article from the same issue entitled: "John Kerry the elusive" bore the subtitle: "Aloof, the Democratic candidate is plagued by contradictions" and lamented his failure to present a coherent impression of himself (Rousselot 2004, 3). Understanding of the election and the candidates was beginning to emerge, however, from the intersection of candidate-based narratives with the *Divided America* and the *Nation in Shock* motifs. Rousselot continued insightfully: "But in a country divided, still placed under the terrorist menace, it is as if all of a sudden the Democratic voters decided that Kerry represented their best chance of dethroning Bush" (2004, 3–4).

Thus, while American conservatives branded Kerry as "French," the French themselves were initially lukewarm in their support for Kerry. The nationalistic sentiment surging in the US was the context in which both presidential candidates framed their rhetoric, and such an environment is by nature hostile to foreign interests or comprehension. Kerry might be better for France, but cutting through his rhetoric to see this fact required more than just reading his speeches and debates in translation; it required a fine sensitivity to American politics that was possessed by only a few French observers. Emmanuel Todd (see Chapter 3) was sufficiently astute to argue that "Kerry's caution should not be misinterpreted. Roosevelt came to power with a classic economic program and he created the New Deal" (Bollaert et al. 2004, 12). Still, as the campaign wore more and more French observers were able to read between the lines. If the socialists in the French government and at *Libération* read between the lines sooner it was perhaps because they were inclined to *write* between lines in a similar way, when communicating their own agenda. In other words, their position in the French political context required a similar deployment of nationalist rhetoric to undercut criticism.

By Election Day, the *Big Difference* motif dominated the French press even though *No Difference* could still be found. Two full-page "profiles" of the candidates were presented in *Le Monde* on November 3rd introduced with the tag: "They are of the same generation and did their studies at the same elite university, Yale, but the two men could not be more different" (*Le Monde* 3 November, 2004, 1). It was now John Kerry, "the fighter," against George Bush, "a ruler with unshakeable certainty who refuses to recognize the smallest error" (Jarreau 2004d, 18; 2004e, 19). The same day on the 4th page, a headline in *Le Monde* read: "The plans of the two candidates reflect two visions of America" (Chambraud 2004, 4) and the newspaper officially endorsed John Kerry in a November first editorial (Le Monde 2004b, 15). The *Divided America* motif was by this point thoroughly imbricated with representations of the two candidates, but the *No Difference* motif still had not disappeared. An article in *Le Figaro* at the end of the campaign (Gélie 2004h) bore the headline, "Two Visions of the United States," but in fact emphasized similarities or presented differences as technical details: a more or less extensive role for NATO in Iraq, a Middle East "initiative" versus a new ambassador to the Middle East, a promise of a new start with Europe versus a promise to move away from unilateralism, a national response plan against terrorism versus a promise to unify intelligence operations, a constitutional amendment to prevent gay marriage versus a promise to support civil unions between homosexuals. On all but the last count, the differences were presented dryly, technically, and with no interest in drawing inferences that might suggest a *Big Difference*.

The French followed the American campaign closely, and as a result they gradually recognized the fact that in the post 9–11 political culture all effective communications in the US must be couched in nationalistic rhetoric. Taking account of this fact, they slowly became aware of John Kerry as an ally, perhaps much more internationalist than he appeared, and they shifted from "anyone but Bush" and *No Difference* perception, toward pro-Kerry and *Big Difference* perception.

A useful threat

A threat to one's nation is generally seen in a negative light. This banal observation is complicated, however, if the threat can serve as a rallying point for nationalist sentiments, and especially if it is not too grave or too immediate. Enemies are useful to the state and to nationalists if they can be used to justify state aims such as military expansion or internal surveillance. Red scares served American nationalists in this way during the interwar period and during the Cold War, anti-Semitism served German nationalist interests in this way during the Third Reich, and Palestinian attacks have served those promoting expansionist policies in the state of Israel. "Enemies of the state" create justifications for the solidification, centralization, or expansion of state power. Islamist terrorist organizations have served this function for right wing nationalists in the United States since September 11 and George W. Bush has served this same "useful threat" function for certain political players in France. In particular for those parties (on both the right and the left) who are seeking to strengthen Europe or create *Europe-puissance,* and who have seen George Bush as a boon because he effectively eliminates the "Atlantic" option. Since cooperation requires two sides willing to cooperate, the Atlantic option is on the back burner until the end of the Bush administration. Non-cooperation, in turn, helps French socialists and others on the far Left and moderate Left, as well as nationalists and those on the Right. For Europeanists, it generates an unexpected momentum. In the words of *Le Figaro*'s François Hauter (2004):

> Whether [the winner] is Bush or Kerry, the Paris-Berlin axis serves as the spinal column of the European Union, and will in any case be put to the test by the result of this American election. Over the past three years it has found a new vigor largely because the Bush administration has achieved something extraordinary, that is to say it has begun to quarrel with its great ally on the European continent, Germany. … If all the Europeans felt themselves to be Americans on the day after September 11, then all the way to England they are [now] hostile to an America felt to be intolerably arrogant.

Embedded in an article about the European Council meeting, which took place in Brussels immediately after the US election, one sees a rather striking political cartoon (Figure 5.3). "We must accelerate the construction of Europe. Bush is only in office for four [more] years," says an EU representative. French opposition to Bush from across the political spectrum ignored his usefulness and chose instead to judge him from the viewpoint of subjects or citizens. The ability of this concern

Figure 5.3 **The text reads: "We must accelerate the construction of Europe. Bush is only in office for four [more] years," Dennis Pessin. Source: *Le Monde* 6 November 2004, p. 8.**

to override the "realist" concerns highlighted in the cartoon clearly indicates the globalization of politics.

Comparison of Sources

In many ways the three newspapers are distinct in terms of format and approach. Some interesting insights emerge by looking in turn at the peculiar features of each newspaper's coverage of the US election. *Le Monde*'s coverage was distinguished by a set of interesting personal portraits linking voters and places to political stances. The other two newspapers did not give the same level of specificity to the faces and places behind the American electoral process. The personalization and emplacement of politics in *Le Monde* worked against anti-Americanism although it may have promoted regional stereotypes. *Le Figaro* excelled in the use of thematic maps to reveal the spatial dimension of the campaign and America's political landscape, while *Libération* was restrained in its use of maps and *Le Monde* came in between. *Libération* was open regarding its bias towards Kerry, and this coverage became quite sympathetic towards half of the American electorate.

Le Monde

Many pains have been taken in media studies to demonstrate that no representation is ever neutral. Bias is always presumed. So it is very difficult to specify exactly what is meant by "neutral" at this point (even when a news source is so neutral as to lose the interest of most audiences, like public radio and television in the US). If it is still possible, however, to use the term "neutral," at least in a comparative way, *Le Monde* provides an example of journalistic writing that is unusually neutral: scholarly, detached, painstaking, and intent on avoiding overt bias. A sense of *Le Monde's* approach is best conveyed by pointing out the few *exceptions* to this neutrality, all occurring in the final days of the campaign. These indications of bias are quite constrained, particularly considering the low stakes involved in taking a stand on a foreign election and the massive preference for Kerry in France. The avoidance of bias probably had less to do with avoiding offense to the portion (albeit small) of the French population that was supportive of Bush, than with maintaining an idealized French conception of the transcendent role of journalism.

Not until October 13th did indications of bias appear in *Le Monde*'s coverage of the election. An article of over 850 words regarding US health insurance concluded with: "the incumbent accuses his adversary of wanting to place health insurance in the hands of the federal government. This is not true, but it is language that Republican voters like to hear" (Jarreau 2004b, 5). The effect was jarring (for those who had followed *Le Monde* carefully) because this sort of language was absent prior to the final weeks of the campaign. The change was evident in a few other fleeting signals. The next day an article appeared that addressed the US conflict over the right to abortion. Although studiously neutral throughout, the very end of this article again indicated a pro-Kerry position. Here the mechanism conveying bias was allowing a particular opinion to have the final word, in this case, a representative of a family planning clinic: "for some we are the last resort… if we disappear, where will people go?" (Dumay 2004, 4). Neutrality is maintained elsewhere in the article by the use of the passive voice (for example writing that the Bush administration policy of promoting sexual abstinence "is critiqued") and by adhering to the story about the US's internal struggle over these issues rather than discussing the issues in general. Only the decision to end with the quote from the family planning worker gives away the bias of the author. In short, *Le Monde*'s policy of studied neutrality almost but not quite neutralized the motif of *The Candidates*.

The clearest and most direct demonstration of bias appeared in the weekend edition of *Le Monde* immediately prior to the American election. Here the newspaper officially endorsed Kerry while acknowledging the extraordinary nature of such an endorsement. "To take sides in a foreign election is not in the tradition of *Le Monde*. The exceptional stakes in the presidential election of November 2, however, and the terms in which this historic choice presents itself, have convinced us that the victory of John Kerry would be desirable beyond the borders of the United States" (*Le Monde* 2004b, 15). The language is ponderous, as if working to regain its composure. The reasons the editors offer for this departure are that Bush is ready to sacrifice the rule of law within America and that he ignores the "international architecture that has been at the center of a world consensus for more than half a century," while Kerry, to his credit, was apparently capable of recognizing his mistakes and had demonstrated the force of his convictions during the presidential debates (*Le Monde* 2004b, 15). The editorial statement concluded that a Kerry victory was preferable: "so that a team is set up in the White House that is no longer guided by Good and Evil, but by law and justice." The newspaper waited until the eve of the election to state its preference in no uncertain terms, but prior to that few hints were given of its position.

In a remarkable series of seventeen short articles appearing generally on the third page from mid October through the first week of November, *Le Monde*'s Patrick Artinian gave a face and a voice to the American electorate. His portraits of American voters intimately explored a wide range of social segments and geographical origins. Each of the short articles included three graphic elements: a photograph of a featured voter, a map marking his or her place of residence in the US, and a 3 x 5 photograph of a place emblematic of the featured voter, such as a store, a factory, a beach, a street, or a ranch. Each profile also included a brief quote from the selected voter and a brief, artfully written description of his or her home town. The portraits, in short, combined photography, cartography, and writing to capture the essence of a political attitude grounded in place.

The people selected for the portraits appeared to be fruit of their environments, and each of them was chosen to reinforce a particular *spatialization* of American politics and a particular *politicization* of place. Carolyn Wiewel's home was Dallas, with: "its skyscrapers, its offices, its city center open to the wind. In the daytime one takes care of business in the proud headquarters of oil corporations and insurance

« JE SUIS À 100 % "PRO-LIFE" : JE VOTE BUSH » – DALLAS, TEXAS

PAR PATRICK ARTINIAN

CONTACT PRESS IMAGES

TEXAS

Election 2000 :
• Bush 59,3 % [47,87 %]
• Gore 37,98 % [48,38 %]
Population : 22,1 millions hab. [290 millions hab.]
Revenu par habitant : 27 752 $ / an [29 469 $ / an]

Dallas

[ ] données nationales

CAROLINE WIEWEL, sans profession, âge secret. *« Je vote pour le président Bush pour une seule et unique raison : je suis à 100 % "pro-life"* [anti-avortement] *et je sais que lui non plus ne supporte pas que l'on tue des petits bébés. »*
GRATTE-CIEL. Dallas, ses gratte-ciel, ses bureaux, son centre-ville ouvert à tous les vents. Le jour, on s'affaire autour des sièges orgueilleux des compagnies pétrolières et des sociétés d'assurances. Après 18 heures, chacun rentre chez soi. Ici, comme dans de nombreuses grandes villes américaines, on habite en banlieue. La nuit, le centre déserté est abandonné aux plus démunis, aux *homeless* (sans-domicile fixe), le plus souvent des Africains-Américains.

Figure 5.4 Voter portrait of Caroline Wiewel from linking person and place. The heading reads: "I am 100% pro-life. I'm voting for Bush." The photo on the right shows an unnamed black woman in downtown Dallas to suggest the segregated character of the city.
Source: *Le Monde* 23 October 2004, p. 3, with permission.

companies. After six one goes home. Here, like in most large American cities, everyone lives in the suburbs. The center is abandoned at night to the poorest, the homeless, and most often the African-Americans" (Artinian 2004d). This is the context in which Caroline Wiewel lives and votes Republican and it is presented as a means of putting that vote in context, explaining it. Her conservative voting behavior is presented as a clue to the meaning of the place, as well (fig. 5.4).

Another portrait shows the other side of Texas political culture. Bart Willis' Democratic vote becomes a clue to the meaning of Austin, in a voter profile that describes the capital of Texas in astonishing terms.

> Right in the middle of Texas, Austin, the capital of the state, gives the impression of a *village gaulois* [Gallic village]. The city where George Bush sat as governor has never devoted itself to the Bush family. Artists, musicians, and intellectuals rub shoulders and give the city a side that is young, dynamic, almost rebellious. Over the years, Austin has established itself as the rising star of country music and rock, and as the symbol of a certain form of counter-culture.
>
> (Artinian 2004c)

The term *gaulois* here is not an ethnic label but rather indicates stubborn resistance. The Gauls – famous for resisting the Roman Empire and featured in the most popular of French cartoons, Astérix le Gaulois – are heroic national figures. Struggling against

« CETTE FOIS, J'ASSURE ET JE VOTE KERRY » – AUSTIN, TEXAS

PAR PATRICK ARTINIAN

CONTACT PRESS IMAGES

TEXAS

Election 2000 :
• Bush 59,3 % [47,87 %]
• Gore 37,98 % [48,38 %]
Population : 22,1 millions hab. [290 millions hab.]
Revenu par habitant : 27 752 $ / an [29 469 $ / an]
[ ] données nationales

BART WILLIS, 38 ans, propriétaire du salon Southside Tatoo à Austin : *« Lors des dernières élections, j'ai voté Nader et je m'en suis mordu les doigts. Alors, cette fois, j'assure et je vote Kerry. Tout ce que m'a apporté Bush, c'est que ma femme vient de se faire virer de son boulot. Il a ruiné l'économie. »*
VILLAGE GAULOIS. En plein cœur du Texas, Austin, la capitale de l'Etat, fait figure de village gaulois. La ville où George Bush a siégé comme gouverneur ne s'est jamais donnée à la famille Bush. Artistes, musiciens, intellectuels s'y côtoient et donnent à la ville un côté jeune, dynamique, presque rebelle. Au fil des années, Austin s'est imposée comme la scène montante de la musique country et du rock, et comme le symbole d'une certaine forme de contre-culture.

Figure 5.5 Voter portrait of Bart Willis from linking person and place. The photo shows Willis' tattoo parlor in bohemian South Austin and the heading reads: "This time I'm goiong to get it done; I'm voting for Kerry."
Source: *Le Monde* 23 October 2004, p. 3, with permission.

tyrannical Romans, they are personifications of a key aspect of national pride, similar in this regard to workers in Russia or cowboys in America (Bernstein 1990, 122). Austin becomes a place of legendary political resistance, suggesting a powerful link of shared interest between France and the American left (fig. 5.5). Place stands in for political identity, and the spatial motif suggests cross-cultural similarities instead of fuelling anti-American sentiment.

American voters were chosen for inclusion in these profiles not simply because of their surroundings but also for the sake of providing a range of interests, beliefs, and affiliations. Kerry supporters chosen by Artinian included two women. One was a 42 year old white trade union official from Boston who said, "I am deeply saddened. We really made the wrong choice. For four years Bush will have complete freedom of action and the right wing of the Republican Party will be able to assert its positions. We're continuing to beat down the middle class" (Artinian 2004n). The other was from a 30 year old English teacher who declared bluntly, "I hate Bush. He's stupid. All he has done is make things better for himself and his clique. I never liked Clinton as much as when Bush was in power. I'm not really excited by

Kerry but I don't think I can stand four more years of this administration" (Artinian 2004m). The political choice, if not the motivation, was the same for a 42 year old black male steelworker from Cleveland: "I'm voting for Kerry, because all the jobs have left the US and what's left are jobs paying 8 or 9 dollars an hour. Here, that's not enough to live on" (Artinian 2004k). The concentration of money and power were dominant themes in the portraits of Democrats.

For the most part, the Democratic voices were urbanites, those listed above as well as a sculptor from Lawrence, Kansas, a urologist from Cincinnati, and a Rabbi from New York City (Artinian 2004g, 2004j, 2004l). Their opinions reflected concerns about job creation and poverty; they were committed and often creative people. One, the Rabbi, remained undecided despite being a Democrat. The only rural Kerry supporter was a black Baptist preacher from Elmore City, Oklahoma (Artinian 2004f). His congregation had dwindled to one person, and it appeared from the context of the other voter profiles that his political peers – rural Democrats – had similarly dwindled to the point of no return. Rural isolation was not portrayed as congenial to the Democratic Party.

The Bush supporters carried distinctly different place associations than the Democrats, dominated above all by isolation and often by rurality. They included an agricultural products sales representative from Boswell, Oklahoma, who saw Kerry as a traitor because he questioned the Vietnam War and a lobster fisherman from Provincetown, Massachusetts, who said, "I voted for Bush. I don't know how to explain it; he just did his job for four years" (Artinian 2004e, 2004p). A hotel receptionist from the tiny outpost of Ord, Nebraska declared, "I'm voting for George Bush because he's a Republican, conservative, opposed to abortion and gay marriage, and in favor of the church and traditional marriage. He supports my values" (Artinian 2004h). The commentary following the last quote described this woman and her rancher husband with uncharacteristic sarcasm. They were: "curious about everything. They would like to know if France has a king or a president, if it's a woman or a man and what his or her name is," and added that the couple did not have time to find out the answers because their cattle needed to be vaccinated. In France, it is common to judge people by the questions they ask (Baudry 2004), so their asking of uninformed questions combines with their evident rural isolation and their preference for Bush to mark them as ridiculous or contemptible. Similarly, the man who saw Kerry as a traitor was described as coming from the middle of nowhere (*"au milieu de nulle part"*):

> A few buildings, rarely any cars. Welcome to Boswell, Oklahoma, on Route 70 between Hugo and Durant. In other words, nowhere. To stop there is almost inconceivable. The life of the few inhabitants is centered around the Dixie Café, the inevitable meeting place. Hunters, musicians, old ladies in curlers rub shoulders a bit like in the saloons of Westerns. Only difference: no whisky here, alcohol is prohibited by local law.
>
> (Artinian 2004e)

Prohibition of alcohol carries a connotation of extremism in France, since wine plays such a central role in the French culture and economy, so this account locates the soul of the Republican Party in a place that is remote, uncultured and bizarre, an echo of Buffon and Talleyrand.

Many of the Republican profiles are portrayals of a rural American mindset captured on television (Chapter 5) by the term *l'Amérique profonde*: rural isolation associated with cultural purity but also with ignorance and backwardness. In this way, many faces and places associated with conservative America suggest a single face of politics: a reactionary politics bred of isolation – an America that has voted and continues to vote in ways that project and animate a severely limited scope of geographical awareness. While the majority of the Republican portraits suggest isolation manifested in vast distances and remote rural towns, those of the few urban Republicans take care to note the isolating facets of American urban life – an isolation created and sustained by racial avoidance. Republicans from Dallas, Boston, and Chicago are all shown with unnamed black people in the accompanying photos – young men on a bus, a woman getting out of a taxicab, an old man buying a hotdog – and the accompanying text in two of the three urban Republican portraits refers to problems with violence, drugs, homelessness and poverty among America's urban minorities (Artinian 2004i; 2004d; 2004o). The conservative white urbanites thus are discursively placed in a setting of ethnic and economic fragmentation. The implication is that American conservatism is produced by forms of social isolation in which urban Republicans willingly participate while rural Republicans are the victims of geographical isolation.

Recalling earlier themes, almost all of the Republican portraits reinforce the motifs of *Nation in Shock* and *Fundamentalist Society*. For example, a woman who manages a legal firm in Boston declares: "We are a great nation. We must unite against the enemies of democracy and freedom. We must help other countries fight against oppression. I think Bush is a good leader, capable of carrying out his tasks well because he is guided by God. And I voted for him" (Artinian 2004o).

Undecided voters and Nader supporters were not overlooked in this remarkable collection of voter portraits. A 29 year old Hispanic video artist from San Antonio declared that he would vote for Ralph Nader because, "For me, Bush means the violation of privacy, the severe regulation of public life with the Patriot Act, a society in which the smallest movement is monitored, spied on. I don't feel secure with President Bush. I'll vote for Nader if he's on the ballot in Texas. If not, I'll settle for Kerry" (Artinian 2004b).

The most important contribution of these voter portraits is that they worked against a simple equation of the 2004 election's outcome with "the" American people, culture, or society. In doing so, *Le Monde* undermined anti-Americanism rather than fueling it. It helped make the alien American society a bit more familiar while leaving intact the perception of American cultural peculiarity. This place-based, personalized perspective revealed the American election as the creation of many different kinds of people with divergent interests coming from many *different places* in both a geographical and a political sense. This is precisely the kind of subtlety that tends to get lost in intercontinental communication flows. To confirm this, one need only examine the homogenizing treatment of the French in the news media and in popular American books such as *The French Betrayal of America* (Timmerman 2004) and *Our Oldest Enemy : A History of America's Disastrous Relationship with France* (Miller and Molesky 2004). By personalizing the election, *Le Monde* generated an effective vaccination against such national stereotyping. Of course, the profiles themselves engaged in *regional* stereotyping, particularly in their association of conservatism with isolating environments, but this geographical motif displaced the blame from American voters to their immediate surroundings. Such an up-close and personal rendition of the electorate problematized any reduction of the Bush victory to a decision of *le peuple américain.*

Libération

The journalistic voice of the French Left has been described by an astute observer of France who also speaks with the insight of a journalist, as "a politically independent, generally high-quality, sometimes brilliant but always irreverent, often snide and cutting daily publication" (Bernstein 1990, 130). "*Libé*," as it is often called, delights in "sly puns [and] clever headlines" (Bernstein 1990, 130) but there was nothing sly about the newspaper's preference for John Kerry (fig. 5.6). Headlines like "France will accommodate itself to the unaccommodating Bush" (*La France s'accommoderait de l'incommodant Bush*) (Soulé 2004) and "Bush gets tangled up in his lie" (Riché 2004a) set the tone, while other headlines went into surprising detail: "Bush, a credo of God and the far Right: fixated on the weakness of his father and obsessed with his faith, the President radicalizes his speech" (Riché 2004e). Worries that Kerry might undermine progress toward European unity could do nothing to make Bush look more inviting on account of the vast gulf in worldviews between the French observers and the American Right. It was not self-interest, therefore, but rather political revulsion that tilted the editorial staff toward Kerry. The headline following the first debate between the presidential candidates reveals this visceral reaction: "Bush counts on slogans, Kerry attempts arguments" (*Bush mise sur les slogans, Kerry tente les arguments*) (Riché 2004b). In its failure to reverberate positively outside of the US, Bush's rhetoric did more to shape the re-presentation of his views by the French Left than did the Left's own interest in an end to Atlanticism.

Personal profiles were also used in *Libération* to explore the meaning and causes of the American election. Although such profiles appeared in a much less consistent way than in *Le Monde*, the impending election prompted this approach on November 2, when profiles of two Democrats and two Republicans appeared on pages 7 and 9 (Libération 2004a; 2004b; 2004c; 2004d). A 51 year old steel worker and a 20 year old struggling actress listed personal economic concerns and "the future of the world," respectively, as their reasons for supporting the Democratic candidate. Differentiated by gender and place of residence, these voters suggested diversity among the Democrats. A pair of white, middle aged, evangelical Christian businessmen gave, on the other hand, a very uniform face to Republican voters. One of them was: "trying to attract young people to Jesus through the café [that he ran]. "As a Christian I prefer the values of the Republicans. Their respect for

Libération

Huit jours pour changer d'Amérique

A une semaine de l'élection présidentielle aux Etats-Unis, «Libération» chronique la dernière ligne droite de la campagne. Aujourd'hui : le vote des minorités. Page 2

Figure 5.6 Image of John Kerry from the front page of Libération with the headline "Eight Days to change America."
Source: *Libération* 25 October 2004, p. 1, with permission.

life. I believe the government must play less of a role in our lives. Like all small businessmen, I'm afraid that the first thing Kerry would do is raise taxes." The other Republican portrait presented a bizarre (to the French) combination of pragmatism and religious fatalism explained as follows: "The Republican approach to the environment seems more 'scientific' to him than that of the 'tree hugging' ecologists. In addition, he continues (without laughing), that global warming is not the doing of humankind, to any significant degree, but rather a 'divine phenomenon'." As important as Christian fundamentalism was in producing impassioned support of George Bush, a far broader segment of the population than just middle-aged wealthy, white Christian fundamentalists voted to return the president to the White House. *Libération*'s use of these two individuals was reassuring but misleading. Also worth noting is that these two Republicans are both from Hershey, Pennsylvania, a city that carries connotations linked to Americanization as the cheapening and adulteration of chocolate, and therefore of the decline of culture in general. *Libération* therefore gave a human face to the American vote, but only through a filter that simplified and caricatured American conservatives while giving Democrats substantially more depth and complexity. This was in keeping with *Libération*'s overt bias towards the Democrats. But this would hardly constitute anti-Americanism because its selectivity is none other than that of a partisan Democrat living in the United States and dedicated to a particular partisan vision of their own leadership.

Le Figaro

Le Figaro did not employ voter profiles to any significant degree. It manifested neutrality regarding the election, did not feed anti-Americanism, and did not take the personal approach exemplified by *Le Monde*. Instead, it treated the election as grand theater and rendered it in sophisticated terms, employing subtle and nuanced reporting and excellent journalistic cartography. The absence of voter profiles necessarily restricted the discussion to the interplay of parties and their candidates. But the paper offered virtually every political perspective from strong critique of the Bush administration to a defense of Bush policies, and a similarly broad spectrum of treatments of Kerry. *Le Figaro*'s "Debates and Opinions" page was a remarkable source for powerful views of all kinds. Here politicians like Hubert Védrine, Pierre Lellouche, Donald Rumsfeld, and Madeleine Albright were in the company of scholars like Carlos Fuentes, Dick Howard (SUNY-Stonybrook), Guy Sorman, Alexandre Adler, André Kaspi, and Emmanuel Todd, and the directors of various institutes and think tanks such as Ian Bremmer (World Policy Institute) and Ken Weinstein (Hudson Institute). Diversity of perspective and reflections from across the Atlantic did much to prevent the dominance of anti-American sentiments.

Figure 5.7 "Principal social questions addressed in referendums." Map shows which states are going to vote on amendments banning gay marriage, medical financing, the minimum wage, and medical use of marijuana.
Source: *Le Figaro* 29 October, 2004: 4, with permission.

Perhaps most striking was the writing style. Literary, verging on poetic, while at the same time intent on details, *Le Figaro*'s writing alone attracts readers of all political persuasions.[1]

> The presidential election ended yesterday evening (late in the night in France) after a merciless campaign of 16 months, undecided to the very end and similar in many regards to a military campaign. What remains is an exhausted and bitter country with smoking battlefields, with the winners in a condition hardly better than that of the losers, and a hangover like the day after a drunken brawl or all-night revelry… This morning, if America has the pleasant surprise of having elected a president, the country may be able to calmly put away its banners and its assassin's slogans, reconcile itself with the dynamism of its democracy, and conclude that in politics as in sports there is always a next time. If again, like four years ago, it is held hostage by a muddle-headed ballot count (*l'otage d'un dépouillement cafouilleux*), by persnickety lawyers (*d'avocats pinailleurs*) and contradictory judges, it will not be able to spare itself a revision of its institutions, a two-hundred year old apparatus conceived in haste in the revolutionary era.
>
> (Gélie 2004i)

1 The service of collecting and forwarding this newspaper was undertaken for me by a staunch leftist who nonetheless judged that the news coverage in *Le Figaro* was superior to that of *Libération*.

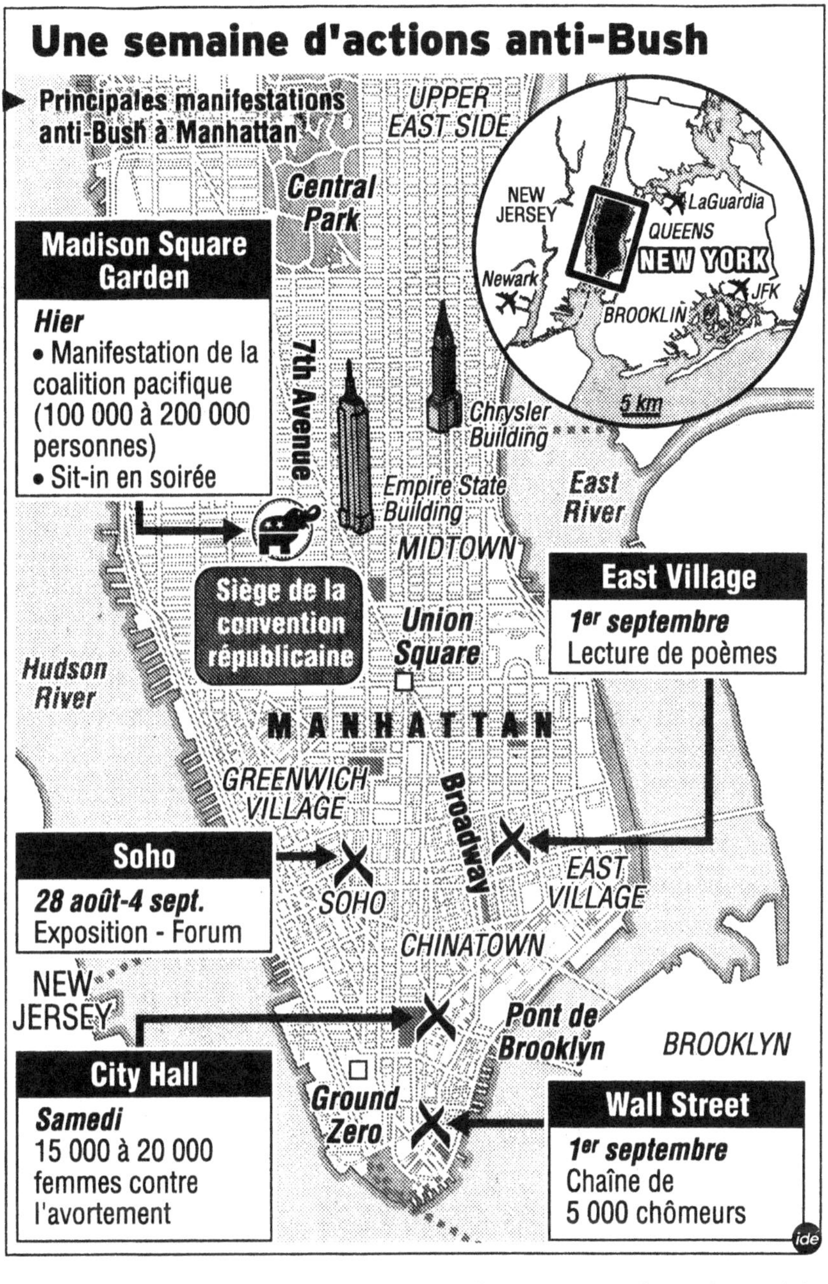

Figure 5.8 "A week of anti-Bush actions" shows the sites of nonviolent demonstrations in Manhattan during the Republican convention, as well as the site of the former World Trade Center. Source: *Le Figaro* 30 August 2004, p. 4, with permission.

The writing evokes a full cast of players, not a homogeneous American character or type. It emphasizes *Divided America*, as well, in perhaps the most striking example of such newspaper coverage. So the reader is positioned above an internal American conflict rather than being urged to take sides for or against a single, monolithic America.

In addition to evocative writing and political neutrality, *Le Figaro* offered readers a rich array of maps. There were maps: distinguishing Kerry states from Bush states (much like those in the American media) (3 November, 3), highlighting the undecided states (30 July, 2; 30 September, 2; 18 October, 2), demonstrating the variety of types of voting equipment used in the US (19 October, 2; 2 November, 2), showing states with contested seats in the Senate (27 October, 2), documenting the regional evolution of the American population and various ethnic groups (26 October, 2; 30–31 October, 4), documenting states with referendums on the ballot (Fig. 4–7, 29 October, 4), indicating economic growth and decline throughout the US (24 October, 2), revealing the variation in political culture within Ohio (21 October, 2), and pointing out the locations of protests in Manhattan during the Republican convention (Fig. 4–8, 30 August, 4). Examples of these maps (Figs. 5.7 and 5.8) show their high graphic quality and detail. Encountering these maps as a collection meted out through time by means of a subscription would have supported a nuanced perception of the US as a space rather than a place – a space internally differentiated at a range of scales by divergent political parties, technologies, issues and perspectives. A map-based awareness of American political diversity would differ significantly from an awareness of American politics based on voter portraits like those in *Le Monde*. But cartographic vision disrupted national stereotypes no less than did personalization and poetic language. In different ways the geographically nuanced awareness of internal differentiation worked against simple anti-American interpretations.

Conclusion

French newspaper coverage of the 2004 election emphasized the idea of America as a geographically, socially, economically and politically divided society. Bush support was thereby situated in space. It was also contextualized in time, as a product of a nation still in shock from the attacks of September 11, 2001. Despite the pervasive French antipathy towards Bush, the primary national newspapers were not immediately pro-Kerry. By the last eight weeks of the campaign, however, the tone in *Libération* shifted to unmitigated support of the Democrats, while *Le Monde* shifted to support Kerry in the last three weeks, particularly in the last week, but was studiously neutral up to that point, and *Le Figaro* manifested a large degree of neutrality until the end. What could account for this neutrality? There was some concern that dealing with a less offensive opponent (i.e. Kerry) might interfere with the project of creating *Europe puissance*, even if the scholarly term was not used. But the appearance of neutrality may owe as much to journalistic and editorial norms, the methods and standards of the newspaper industry, which in France takes objectivity very seriously. The weight of more than two centuries of symbolic constructions of North America and "America" by French intellectuals also argued against the probability of an overnight change under Kerry's hand. More important than the particular man in charge, the unforgettable images of the collapsing towers supplied an explanation for America's political climate with its excessive military posturing from both parties.

Some long-term concerns of the French with regard to American society were treated at length in the stories about the election: ideas of America as a fundamentalist society, an economy tottering on the brink of collapse, and an unchallenged hyperpower striking out at will against enemies real or imagined. The intellectual roots of these positions are demonstrated in chapters 3 and 4. For all of these worries and concerns, the press coverage of the election did not take a uniformly negative view of Americans or American society although on the topic of religion all of the newspapers were outspoken and negative.

Whether by focusing on particular American citizens and their surroundings (*Le Monde*'s specialty), or by mapping social, economic and political diversity (*Figaro*'s specialty), or (in the case of *Libération*) by siding with half of the population, the French newspapers generally contributed to their communication environment a range of ideas that were not compatible with simplistic anti-Americanism. In fact, its focus on personal interest (in the case of *Le Monde*) on geographical distributions of American political culture (in the case of *Le Figaro*) or on the struggle of half of the American electorate to prevent Bush from regaining power (in the case of *Libération*) clearly demonstrated that the outcome of the election was not the will of "the Americans" in an all-encompassing sense. Their contribution to public debate could potentially help construct an informed polity capable of reflecting on the challenges facing democracy and citizenship in the early 21st century. Print journalism was, of course, not ideal in a Habermasian sense or otherwise. It was rooted in French perceptions, conceptions and representations of the US, in other words, in a set of familiar motifs, some drawn from French intellectual thought, but on the whole it was supportive of the continued expansion of polity beyond national borders through mutual recognition and understanding.

On a final note, the decline of newspaper readership in France is a cause for concern. Since the US's overwhelming power makes anti-Americanism virtually inevitable as a motif circulating in international communications in the foreseeable future, the question remains whether the media that are rising to take the place of newspapers in France, the Internet and television in particular, will convey as effective an understanding of American politics. The following chapters permit reflection on that question.

Chapter 6

Television: Plumbing the Depths of *l'Amérique Profonde*

> The Americans never use the word "peasant"; the word is unused because the idea is unknown; the ignorance of primitive times, rural simplicity, and rustic villages have not been preserved with them, and they have no idea of the virtues or the vices or the rude habits and the naïve graces of a newborn civilization.
>
> de Tocqueville, *Democracy in America*, p. 303

In April of 2003, Tomas Kellner of *Forbes* magazine announced French plans to form a cable television news channel in words dripping with sarcasm: "CNN à la Française," he called it. "Much too late to cover the Battle of Bagdad. But still time, perhaps, to roll out a documentary about the long, cozy relationship between Chirac and Saddam Hussein" (Kellner 2003, 52). This channel was, in his view, intended to "push the French point of view on world events," a goal politically motivated by Chirac's "contretemps" with George Bush. But the slowness of the cable news initiative betrayed classic French inefficiency and obstinacy: "Hélas, the real fight lies at home. Although the idea has been around since 1989, it has gone nowhere because France's myriad state-run broadcasters could never agree on funding and ownership. Chirac's revolutionary solution: Engage the private sector" (Kellner 2003, 52). Reinventing the wheel, debating worthless alternatives, mindlessly copying, disseminating propaganda – this is what the French were good at. Behind this sneering article lay assumptions about French culture and French media, television in particular.

Kellner's article recalls the "French-bashing" stereotypes evident in other American media in 2003 and 2004 (see chapter 3). These stereotypes would have us believe that French television could only cover the American election in self-serving, politically-motivated ways. The French television news reports might be expected, for example, to use the thoroughly disliked American president as a symbol of America to help drum up French outrage at the US. This chapter demonstrates, on the contrary, that the majority of French television coverage was remarkably free of overt and even covert forms of anti-American bias. More than that, French television worked against the tendency to equate the disliked Bush administration policies with the US and with Americans in general. Instead, French television constructed motifs that offered viewers an opportunity to sympathize with at least half of the American electorate and to better understand the other half. Like the voter profiles in *Le Monde* (chapter 5), television achieved this in an intriguingly geographical way, by linking political attitudes to the peculiarities of American places. Television's contextualization of the election encouraged French viewers to excuse Americans for their support of Bush in light of their peculiar environments. Although the outcome of the election might be distressing, only some Americans were to blame, and of these misguided Americans were victims of their environments.

Television Affordances

Television's unique affordances shape the presentation of news. Compared to newspaper articles, television news reports contain far fewer words. Coupled with the telegraphic use of words is a visual language that is rich and complex. Television images – moving, colorful, delivered in a carefully planned sequence and pace, juxtaposed in a sequential pastiche – convey very special kinds of information that are impossible to convey in words. As a result, television has a particular relationship to the topics it treats, a "visuality" that is seldom as clearly evident as during a political campaign when politics becomes theatrical. Globalization simply expands this role beyond borders.

The element of motion makes it possible for television to show rather than tell. The World Trade Center attack is a case in point. The television images of this event packed an emotional punch far beyond that of the written accounts. Televisual images surpassed even the arresting still photos of burning skyscrapers because they included dynamism, sequence and pacing: the airplanes gliding towards the towers, the towers with their geometric purity, the planes impossibly close to the towers, then the dissolution of one solid object into another, the bursting fireball, the smoke and shattering glass, faces imprinted with shock and horror. Words can evoke television coverage etched on memory (as they have just done) but they lack the emotional explosiveness of television's moving images. In still photos, the same event becomes an emblem, frozen in time and "overrun" by meanings (Adams 2005, 85–91; Barthes 1967; Barthes 1972; Sontag 1977), yet less immediate and therefore less emotionally compelling. In moving images the event retained the greatest proportion of its "life." Not coincidentally, the term "live" television makes sense whereas neither still photographs nor text can be described as "live." In comparison to written descriptions, a location captured in moving pictures acquires depth and dynamism; it ceases to be simply a location and takes on much of the complexity and particularity geographers associate with the term "place" (Tuan 1977; Adams 1992).

In addition to moving images, television delivers sound. Sound forms another link to place allowing audiences to vicariously experience presence in the unfolding of an event. The screams and exclamations in the videos of the September 11 directed audiences to share in the shock felt by bystanders. Of course no communication medium can convey emotion directly to audiences, yet some media facilitate the transmission of emotion better than others. Telephone captures more of the emotional subtleties of communication than the silent medium of e-mail. Newspapers serve the function of record keeping and the documentation of factual details. Television handles dramatic content well not simply because it conveys action, but also because it enlivens action with sound to produce a *gestalt*: a multi-modal, complexly textured and emotionally involving experience unparalleled by other media.

But what of television coverage of topics that do not involve moving objects and interesting sounds? What about television's political coverage? A political contest lacks the dynamic real-time elements of a terrorist attack. The details of a political speech or debate are much easier to capture and transmit in printed words than in images and sound (although the decisive factor here is not technological but economic). Only in the most unusual cases does television actually devote costly air time to a speech from beginning to end, and only the most dedicated viewers will attend to such coverage. The exceptions, of course, are State of the Union addresses and acceptance speeches at party conventions. But these are rare events and they attract audiences despite the affordances of television rather than because of those affordances. But there is another side of politics – the world of citizens and of *political culture* – that lends itself much better to television. A political rally captured by television cameras combines architecture and decorations, clothing and hair styles, gestures and tone of voice, music and noisemakers, balloons, banners, and swirling confetti. Reduced to words the rally is basically a non-event; speakers say little of importance as they mouth familiar slogans. So campaigns make "good" (read marketable) political television.

Television's gestalt of moving images and sounds is ideal for capturing a sense of place, which also permits political culture to be rendered with a high level of detail. A conservative (or liberal) town may be described on paper but it remains little more than an abstraction. The same town can represent conservatism or liberalism in a much more captivating way on television. Every public gathering place in the town, from churches to cafés to street corners, can convey volumes about the groups that struggle for dominance in the place. Every location can become a synecdoche – a part that stands for the whole: a church stands for conservative voters; a closed factory stands for corporate power. The televisual language creates a pastiche of people, places and objects as the framework for political meaning. To offer another example, a slow pan down a suburban street followed by a close-up on a "Bush/Cheney" sign invites the viewer to make a connection between this place and its political alignment, fusing landscape and ideology. Television breathes life into political abstractions and allows the *placing* of politics.

Much is intrinsically and almost unavoidably communicated about a place by the camera's roving eye as it moves over, around and through the locations of its stories. This is true notwithstanding the fact that a camera also excludes, and can easily create the illusion of showing everything even as it hides major portions of a place (a fact that makes three-sided rooms entirely effective as film and television sets). More generally, what comes across particularly well on television is a *sense of the particular* that may be embodied not only in place but also in personal identity. In regard to personal identity, a speech shown on television reveals much about the speaker's mood, attitudes, physical appearance, way of dressing, way of moving, and so on, while the same speech transcribed in a newspaper includes only the words, or at best a reporter's second hand impressions of visual details. Television's accentuation of the personal identity and place are mutually reinforcing; viewer attention to one adds automatically to the other.

In contrast, what often fails to come across on television is the complexity of issues. Causality, specifically of the interplay of multiple causes or the potential for various outcomes from a single chain of events, is often missing. This limitation can again be traced to television's unique affordances, since an understanding of causality requires the simultaneous consideration of past and present in a logical relationship. Words are uniquely suited to create this awareness (Adams 2005, 80–90; Postman 1985; Kress 2003). Television can certainly explain the causes of things, but it does so in a more cursory fashion than do written texts, and its causal explanations are supplied around the margins of a cascade of evocative and gripping images that draw attention to personalities and places.

The danger in this televisual language of personality and place is that it appears natural, yet as Fiske and Hartley argue (1978, 17) "It is by no means natural for television to represent reality in the way that it does, just as it is by no means natural for [verbal] language to do so." The Bush/Cheney sign juxtaposed through the cutting together of two television shots may in fact express the preference of only half of the people on a block, but as a visual icon the sign stands for the entire block. We do not know if the camera is aimed carefully so as to leave out complicating elements of place, whether antithetical to the interpretation intended (a Kerry/Edwards sign) or simply distracting (a yard sale). The act of selection involved in constructing a television sequence is one reason television is always in some sense political. What is left out is marginalized and its very possibility may be forgotten; what is captured becomes not just an image but a natural, or rather *naturalized,* image. Hence it is a lever by which to manipulate audience sympathies.

When an event threatens pre-established worldviews television must do symbolic work to recuperate familiar worldviews. It consequently attacks the unexpected and counterintuitive elements of experience, seizing the "out of place" phenomena and forcing them back into place, a strategy Fiske and Hartley (1978, 87) refer to as "claw back." The politics involved here is often the mundane politics of daily life whereby the televisual image favors certain actors and perspectives while marginalizing others (Fiske and Hartley 1978, 85–100; Williams 1974, 49–54): focusing on men versus women, attractive people versus unattractive people, white versus nonwhite, young adult versus the elderly and children. But the politics of the image also spill over into formal political spheres, such as electoral contests and international conflicts, where televisual inclusion and exclusion operate in much the same way, rendering certain views and actors in a more positive light while casting others into doubt or simply leaving them unknown and hazy. Klaus Dodds (2000, 24) describes such a link between television and the shaping of political opinion: "Public apathy in the United States and Western Europe towards the bombing of Iraq between January and March 1991 was in part created by television images of empty desert landscapes rather than a complex society composed of 80 million people. The lack of images depicting the fate and effect on the population gave the impression of war as a video game rather than as wholesale destruction." But seeing something in greater detail, or even seeing it with human inhabitants, does not always lead to appreciation or sympathy. Television's sensory qualities can accentuate the sense of cultural distance even as they seem to transcend physical distance; they can create an alienated view of the Other that differs strikingly from McLuhan's "global village" (1962, 31–2).

The appearance and actions of distant people may seem downright bizarre when viewed in great detail or when framed according to a motif that promotes incomprehension. For example, the person selected as spokesperson for a particular group by the televisual gaze may be the most outrageously dressed member of the group. This occurs on television because such "performers" draw viewer interest. A spokesperson may be given only a few seconds to articulate complex ideas, on the assumption that people are not interested in the details of an argument they have not previously heard. The opposing spokesperson may be an expert with a degree, and he or she may be given more time, or the first word, or the last. The favored spokesperson may be better lit, filmed in a better environment, or filmed with certain "props" such as a petition or court order. In these and countless other ways a televisual text may favor a particular political stance (Gitlin 1980). In composing a news segment, the television reporters, editors and photographers may be unaware of promoting a particular bias, but that does not in any way mitigate the political effect of their work.

One way to understand the effect on viewers is in terms of audience "position." The various elements that make up the television text can be strung together in the mind of the viewer in various ways, but some of them make more sense than others. The televisual text is like a kind of puzzle that only holds together when the pieces are assembled as intended. Therefore, television "produces a socially located position that it invites the viewer to occupy in order to understand it easily and unproblematically" (Fiske 1987, 25). By "occupying" the intended subject position in relation to the text, the viewer discovers that the text makes sense in all its details, and this intended subject position is a way of reading the text that offers an illusion of knowledge, and hence power, so its force is not insignificant. The construction of a text to encourage a single reading is a hallmark of textual "realism," and the constructedness of the text means that textual realism is always a kind of deception. Thus the news, through narrative realism and live imagery, constructs the world and the audience according to implicit political biases while disguising that construction as a natural or transparent view of the world.

The presence of sound in television texts multiplies the political power of its juxtaposed and sequential images. We can imagine a pan of a suburban street accompanied by different soundtracks: the "voice over" of one candidate or the other, an interview with a worker or the sound of voices singing a hymn, the sound of children playing or an ominous silence. The choice of what sound to overlay the moving images radically affects how audiences read a televisual representation of place, and in turn how they respond to the implicit political messages. When translation is necessary, as in news coverage of speeches given in a foreign language, not only is there some latitude for interpretation of word meanings, but the ability to shape the voice quality and intonation of the dubbed version also impose yet another a layer of constructedness between the events and the viewers.

But television's peculiarities as a news medium derive from more than just these technologically-defined affordances of the medium. Television, like any technology, is appropriated within a social context that favors certain uses and discourages others. Generally speaking, as a social institution television is crafted by "authors" with a strong commitment to central or mainstream perspectives. This conservatism can be attributed to funding sources, its relatively sophisticated infrastructure, and its social status as a "bardic" medium (Fiske and Hartley 1978). When television presents politically "peripheral" perspectives it is generally on less popular channels, on public channels like PBS in the United States, and on cable channels. But the high cost of television production and the competition for major audience segments leads to an avoidance of controversial positions by commercial channels. Furthermore, television's naturalness and "realism" also create a bias towards a priori assumptions that permit easy interpretation: "Realism involves a fidelity both to the physical, sensually perceived details of the external world, and to the values of the dominant ideology" (Fiske 1987, 36). This does not mean that controversial positions are entirely absent on television, but they are usually presented within an overall narrative framework that works to neutralize their radical potential (Fiske 1987; Gitlin 1986; Williams 1974). Television is therefore situated within social processes as a central institution that recuperates marginal viewpoints to build a centrist "common sense" view of the world, even as it leaves some opportunity for alternative or "deviant" interpretations (Adams 1992).

The political power of texts is seldom recognized by those who are unpracticed in creating such texts, and the vast majority of any audience falls into this category with regard to television. Thus while most people can understand television sequences by early childhood without any formal training, they remain functionally illiterate in that they cannot "speak" the language of television to articulate their own worldviews. They can physically point a videocamera but they do not understand the most basic techniques of cutting and juxtaposition, let alone the subtleties of creating a subject position or realist television narrative. This televisual illiteracy increases television's power to naturalize mainstream views by ensuring that its constructions are difficult to deconstruct. To the viewer one party or the other in a conflict simply *seems* more persuasive, powerful, or competent. The judgment seems to remain in the hands of the viewer rather than in the hands of the video producer, and the political judgment seems natural, given and real rather than humanly constructed. In terms of affordances, then, television does more than just capture life through moving pictures and sound. It employs sights and sounds orchestrated through juxtaposition, sequence and pacing to create a gestalt in which political biases are conveyed in an imperceptible yet powerful way.

The cycle between intended understandings and derived meanings is never closed, however. People draw on other experiences, both direct and mediated, to decipher and interpret televisual experiences. Whereas all communications involve some degree of "slippage" between the intended reading and the plurality of interpretations, television is particularly "open" to divergent interpretive frameworks (Fiske 1987). We cannot easily acquire knowledge about the interpretive frameworks of French viewers. Nonetheless, in light of the apparent naturalness of the television language, the power of an audio-visual gestalt, and the audience's "illiteracy" in that language, it is not unreasonable to assume that its motifs are widely "read" in ways that perpetuate the biases most evident in the frames. In the language previously introduced, the media frames of interest are geopolitical motifs. Following a brief "teaser" introducing the way bias appears in television texts, and an introduction to TF1 and France 2, our two sources, we will turn to the motifs employed by television coverage of the American presidential election in the fall of 2004.

Teaser

Consider the following segment from France's TF1 (the main commercial station) on the night of October 6, 2004. A report on the Edwards-Cheney debate followed a story about a car bombing in Falluja, Iraq, which killed sixteen people and wounded twenty. The debate report was, in turn, followed by a story on Jacques Chirac's call for a global moratorium on executions. Not only was the Edwards-Cheney debate sandwiched between stories that reflected negatively on the US (and in particular on key policies of the Republican administration) but the points of transition between the three news stories emphasized connections that were un-flattering for the US. The transition from the car bombing to the vice presidential debate was facilitated by the news anchor's announcement: "Iraq was at the heart of a lively exchange of fire (*passe d'armes*) yesterday in the vice presidential debate." Armed struggle in Iraq offered insight into the political struggle in the US, particularly since the Democratic candidate was able to benefit from a recently released report that "buried" the Bush administration claims regarding weapons of mass destruction in Iraq. Footage of John Edwards demanding honesty from the Bush administration and criticizing the administration's incoherence was shown next, dubbed into French, followed by a much shorter quote from Dick Cheney insisting that no mistakes had been made in Iraq, a reply framed as: "An obstinacy that bears fruit in public opinion. At least up to this point" (TF1, Oct. 6).[1] This story was followed, in turn, by Chirac's call for a moratorium on executions and a subsequent report on the death penalty around the world, which was again linked to the campaign. "Seeing as how we are speaking about the United States, Jacques Chirac called yesterday for a global moratorium on capital punishment" (TF1, Oct. 6). This link was, in a sense, "empty" of content, yet the logic behind the link emerged from the capital punishment story, which showed the United States, Iran and China on a map as the countries with the most executions, then presented an interview with a female inmate on death row in the US, and showed stark images of the rooms and equipment used to carry out capital punishment. The sequence and linking of these three stories resonated with

1 Citations of the 8:00 pm news on TF1 will be listed simply as "TF1" with the corresponding date; similar notation will be used for news reports on France 2.

French critiques of American foreign policy, law enforcement, private weapons and punishment. Through television's grammar of sequential juxtaposition, the US under George Bush was linked by causality and similarity to China, Iran, and Iraq, and more generally to places of chaos and a geography of violence.

To what degree was this kind of symbolic juxtaposition the norm in coverage of the 2004 election? The question is difficult to answer because the main international news stories of fall 2004 in France were the war in Iraq and the American election. The "international news" segment within each night's program would automatically group the American election with the war in Iraq. Looking throughout the election coverage on two television channels, it appears that no consistent effort was made to discredit the US through juxtaposition. The sequence presented above was highly unusual. Furthermore, a careful look at the motifs employed by French television to represent the American presidential campaign shows how the *placed* and *personal* quality of television news coverage generally discouraged an anti-American reading of events. The interesting story behind the French television news was not anti-Americanism but rather how political culture was *localized* or *placed* within the US. Just as in the newspaper certain locations became emblematic of support for Bush, while at the same time the US as a whole was revealed as deeply divided, with some regions quite passionately opposed to Bush.

The Sources

Télé-France 1 (TF1) and France 2 (FR2) are the two main broadcast channels in France. Together they capture about 60% of the French broadcast television audience (35% for TF1, 25% for France 2) and supply virtually all of the international news coverage. While France 2 is a public channel and TF1 is commercial, the distinction between public and private broadcasting is much less noticeable in France than in the US. Commercial channels are allowed to interrupt a program only once during the course of an hour, so the ads come mainly between programs like the informational announcements on public television. Both types of channels are politically independent in France and at times express dissenting political views. Both public and private television at times question, critique, and lampoon government officials. The main difference is that public channels depend on government funding (Bertrand 1999, 80). The interests are different, however, since the private network ultimately *produces* an audience – a commodity for resale to advertisers – while the public network *serves* an audience as citizens. For TF1 the audience is a means to an end whereas for France 2 the audience is the end.

Sold to the "cement king" Francis Bouygues, when it was privatized by the conservative Georges Pompidou government in 1987, TF1 currently shows annual earnings close to 400 million Euros (Bernstein 1990, 192; Bouygues 2006). The acquisition joins four other branches of the Bouygues Group: Bouygues Construction (large construction projects), the Colas Group (roads and transportation), Bouygues Immobilier (real estate and property development), and Bouygues Telecom (telecommunications). With over 23 billion in sales in 2004 and operations in 80 countries, the Bouygues Group necessarily approaches television from a different perspective than France 2, directed by France Télévisions, under the restricted regulatory authority of the federal administrative body called the CSA (Conseil Supérieur de l'Audiovisuel).[2]

Not surprisingly, public and private television in France handled news of the American election rather differently. We will see that the public channel provided more in-depth, extended coverage of the election and the state of the US in the election year, while taking more political risks than the commercial channel. In this regard, its role was much like that of public television in the US although the contrast between public and private was somewhat less pronounced than in the US.

This study includes all material related to the US election on these two channels but focuses in particular on the 8:00 news. Eight o'clock pm (known locally as 20h00) is the time of the most in-depth televised news programs on both channels. In this time slot the two channels account for exactly two thirds of the total television viewing audience (40% going to TF1 and 26% to FR2). This degree of attention indicates that an in-depth study of the 8:00 news on these two channels provides a fairly complete sample of the motifs employed on French television news to represent the American election. In addition the study includes all other program content related to the US from July through October 2004, utilizing the computer subject index provided by Inathèque at the Bibliothèque National de France.

The two channels were initially very similar in coverage with close to two minutes of coverage each night during the key events of the campaign (the Democratic and Republican conventions and the three debates), but during the late phase of the campaign October 11 through November 4, FR2 devoted considerably more time to the campaign during the news, reaching a high of 40 minutes on November 3, while TF1 allocated only 23 ½ minutes to the event that night. Non-news programming related to the election showed a similar difference, as FR2 clearly established a pattern of interest throughout October. The night after the election, France 2 again demonstrated its peculiarly strong commitment to the story by airing a post-election report from Medina Ohio, which was presented as a further demonstration of the political climate in *l'Amérique Profonde* (Deep America), a generic place motif that became a preoccupation during the television coverage.

Table 6.1 shows the duration of coverage on selected days before and during the election. Figures 6.1 and 6.2 show the entire volume of coverage of US-related

---

2 According to the website of the CSA, "The Conseil supérieur de l'audiovisuel is an independent administrative authority that was created by the Law of January 17th, 1989 to guarantee broadcasting freedom in the conditions laid down by the modified Law of September 30th, 1986. Nine Conseillers are nominated for a period of six years by presidential decree. Three of these members including the President are designated by the French President, three by the President of the Senate, and three by the President of the National Assembly. Three of the mandates are renewed every two years and the functions of the members of the Conseil are incompatible with any other term of office, the civil service or any other professional activity." http://www.csa.fr/multi/introduction/intro.php?l=uk.

material on the news and in other television programs. On TF1 the total news coverage from the beginning of July through November was close to three hours (2 hours, 54 minutes and 34 seconds). On France 2, the total news coverage during this period was 25 percent higher (3 hours, 56 minutes and 13 seconds), confirming the public channel's pattern of slightly greater attention to the US election.

Not surprisingly, the 8:00 news programs on both TF1 and France 2 showed a sharp spike in coverage of the campaign in late October and early November (figures 6.1 & 6.2). TF1 reached a maximum weekly total of just over 84 minutes during election week, while France 2 exceeded that figure by 40 percent. The only election-related story longer than four minutes prior to mid October was a TF1 interview with President Clinton on July 15. The longest coverage on a single night was 28 minutes on TF1 (Nov. 2) and 40 minutes on France 2 (Nov. 3). Small spikes in interest corresponded with the Democratic and Republican conventions, both channels giving more time to the Democratic convention than to the Republican convention, in part because of the presence of a significant contingent of French politicians at the Democratic convention that was either not present or not reported in the case of the Republican convention. Other small spikes followed each of the three debates then, during the week before the election, stories from foreign correspondents stationed throughout the US began to appear with regularity. While coverage durations of the two channels were rather similar throughout most of the campaign, France 2

Table 6.1 Time dedicated to key events in the 2004 US presidential election by France 2 and TF1

Event	Date of broadcast	duration in minutes: TF1	duration in minutes: FR2
Democratic Convention	July 27	3:00	1:45
Democratic Convention	July 28	2:00	0:22
Democratic Convention	July 29	1:55	2:10
Democratic Convention	July 30	2:00	2:52
Republican Convention	Aug 31	1:30	2:15
Republican Convention	Sep 1	1:45	2:00
Republican Convention	Sep 2	0	2:15
Republican Convention	Sep 3	2:00	2:00
Debate: Bush vs. Kerry	Oct 1	4:00	2:36
Debate: Cheney vs. Edwards	Oct 6	1:44	0:20
Debate: Bush vs. Kerry	Oct 14/15	2:00	2:30
Election	Nov 1	22:45	23:00
Election	Nov 2	28:41	37:30
Election	Nov 3	23:21	41:00
Election	Nov 4	3:30	9:25

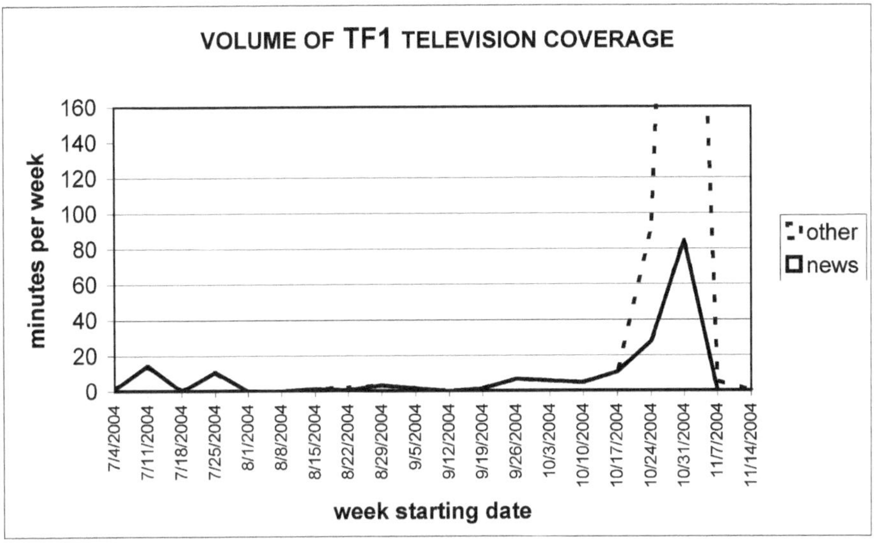

Figure 6.1 **Volume of coverage of issues relating to the American presidential election, the candidates, the parties, and American culture and society in general, on the commercial channel, TF1, between early July and mid November.**

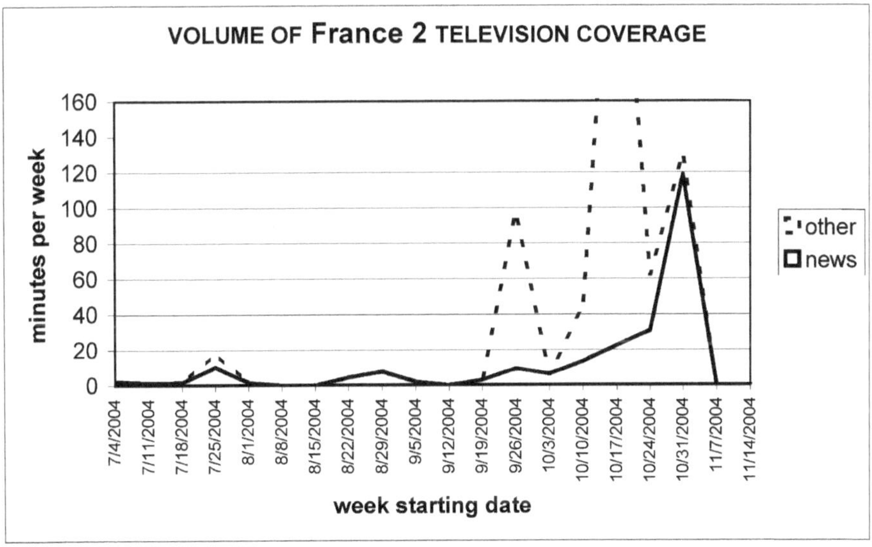

Figure 6.2 **Volume of coverage of issues relating to the American presidential election, the candidates, the parties, and American culture and society in general, on the public channel, France 2, between early July and mid November.**

diverged noticeably at the end, with 30% more coverage on November 2, 74% more on November 3, and 169% more on November 4.

The picture becomes a bit more complicated when we include non-news programs, such as special reports and talk shows. The commercial channel, TF1, appeared to devote more time to the election and related issues outside the 20h news, with 8.6 hours as compared to 7.6 hours for France 2, but over 80 percent of TF1's non-news time was logged on November 2 and 3, owing to all-night live coverage of the election despite the fact that there was little real information to report during the night. France 2 distributed its non-news US-related coverage over the month and a half leading up to the election in the form of a flurry of special reports, discussions on talk shows, and special interviews, many of which were at convenient viewing times and addressed substantive political and cultural issues. The public channel thereby demonstrated particular interest in examining the roots of American political culture, especially neo-conservatism. The week of October 23rd brought almost five hours of such American culture material. A special one hour and forty-five minute edition of the documentary program *Un œil sur la planète* (An Eye on the Planet) that aired during this week gave a wonderfully in-depth, if rather idiosyncratic, view of contemporary America. The talk program *100 minutes pour comprendre* (100 Minutes to Understand) went well over 100 minutes, as well, providing a discussion forum for intellectuals of statecraft from both countries including Zbigniew Brzezinski and Hubert Védrine. At least some notice was taken of the election across a broad spectrum of France 2 programs, including even a religious program and a sports program (24 October and 31 October 31).

Another important measure of interest is placement of stories within the time interval of the news program. Figures 6.3a and 6.3b show the placement and duration of election news coverage within the two 8:00pm news programs. Obviously, more important stories, in the judgment of news producers, are placed earlier in television broadcasts. Start times for news of the election differed between the two channels. Election coverage appeared closer and closer to 8:00 pm on France 2 as the date of the election approached, sliding from around 8:25 to 8:15 by the end of October. TF1 treated the lead-up very differently; the commercial channel began regularly placing election stories between 5 and 10 minutes into the program starting in September, which helped build the election as a news focus and establish an audience dedicated to following the story. Eight stories were foregrounded in the 5–10 minute time slot although on a dozen different occasions stories were scheduled much later in the program. Public television fulfilled its role as a source of information and debate at the heart of the public sphere, addressing the event in several different forums, and developing a deep and sustained analysis of the election with news stories related to the election on three quarters of the nights between September 20 and the end of October. Commercial television treated the election in a more episodic fashion, spending less time overall but strategically building interest in the story with eight short news segments placed within the first 10 minutes of the 8:00 news program between September 19 and the end of October. What is indicated is an interest in capitalizing on "attention-grabbing" events by placing them early in the program

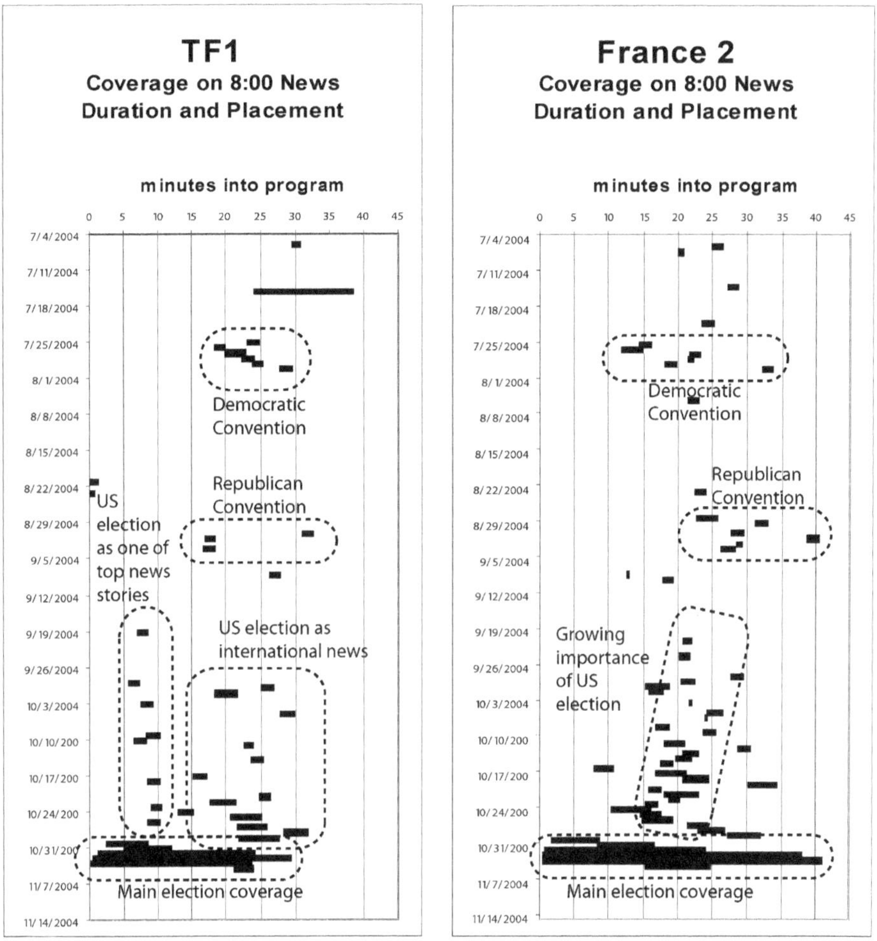

Figures 6.3a and 6.3b **Coverage of issues related to the US election on TF1 (commercial television) and FRANCE 2 (public television), by day and time during the 8:00 news.**

and thereby making the news more captivating as a commodity while keeping the minutes of costly foreign reporting relatively low.

Starting in early November the election news reports occupied substantial portions of the respective news programs. The observed durations of over 20 minutes were striking for news of a foreign election, particularly in light of the fact that the 8:00 news in France generally lasts between 40 and 50 minutes rather than a full hour. At this point the election climbed to the top of the nightly news with start times at or very near to 8:00 on both channels for three days in a row. After November 3, news of the US faded away quickly. The post-election analysis fell back to the "international" time slot after the fifteen minute mark for one day then disappeared altogether. In short, the shifting placement of election coverage within the context of the 8:00 news throughout the study period suggests a long period of rather mild interest followed by brief fascination then rapid loss of interest on both channels.

Thus, to slightly different degrees, both of the television stations conveyed the extraordinary nature of the American election but their time allocation and scheduling of stories within the hour betrayed the difference between commercial and public television. Both stations allowed the election to become a major preoccupation for at least a week. Both stations gave the election a degree of attention unimaginable were it to have occurred in any country other than the United States.

These quantitative findings point in general to a televisual construction of the US election as very important news, on both channels, albeit handled in different ways. This finding is confirmed by the nature of the coverage in terms of its subject matter and motifs, as well as the verbal explanations given by on-site reporters and news anchors. All indications from the television news point to the extraordinary importance of the American election to the French. We turn now to the set of motifs that constructed the American election for French television audiences.

The Motifs

Importance

"American election but global stakes," declared France 2 news anchor David Pujadas (France 2, Nov. 3) to encapsulate the significance of the election. This verbal framing mirrored the station's decision to place the news anchor overseas. On November 1, both of the stations signaled the election's importance by relocating their news anchors, along with a crew of reporters, to the US in order to document the election at close proximity. The use of overseas anchors gave immediacy to the event and New York was an emblematic location. Both Patrick Poivre d'Arvor and David Pujadas were filmed in front of the New York skyline. Pujadas was positioned so that viewers could observe the movement of traffic on Broadway while in Poivre d'Arvor's case the camera was oriented to capture the words "New Yorker" at the top of a facing building. TF1 scattered over a dozen reporters throughout the US – in New York, Washington D.C., New Hampshire, Massachusetts, Florida, Tennessee, Texas, and Ohio. France 2 was a bit more sparing with its reporters, but more generous with the time it afforded to special reports from the emblematic small towns and rural communities that served as icons of America's political heartland. Paralleling the geographical shift to an American observation site in November was a shift in the timing of stories about the election to the top of the news hierarchy at the beginning of the news program. Furthermore, the duration of US-related news stories mushroomed to a half an hour or more each night. A concerted effort was made by both channels to capture the mood in American cities and towns on the occasion of the election.

Various indicators – geographical placement of reporters and anchors, temporal placement of the election within the news, as well as total time dedicated to

American news – all indicated that the US election was judged by French television in its gatekeeper role to be a very important historical event. These indications of importance were confirmed by Patrick Poivre d'Arvor as he launched his on-site coverage with the following explanation:

> We are in the United States for these first three days of November because the election that will unfold here tomorrow is of major importance. Paradoxically, the man who will be elected … will not be the most powerful person on the planet, or in the United States, in any case, because the financiers and industry leaders are well ahead here and also because the Congress is a non-negligible counter-balance, but he will be incontestably the most powerful person in the world as the recent events in Afghanistan and in Iraq have amply demonstrated.
>
> (TF1 Nov. 1)

Although ambiguous, this statement does reveal the French judgment of American society as dominated by capitalists, a long-standing French critique of the US (see chapter 3 and 4). It also subtly points out the tension between France and the US over the invasion of Iraq, a subject topmost in the minds of many viewers. Thus Poivre d'Arvor not only situated himself in the US with this prologue, but he also situated the audience inside the dominant French worldviews.

Not inconsequential in television news reporting is the effect of the personality projected by the news anchor. Poivre d'Arvor is a prolific author and the host of two talk shows relating to literature. At the age of 58, with a serious and slightly stodgy manner, he evidently would appeal to older and more intellectual audiences. The competing reports from France 2 were presented by David Pujadas. Younger than Poivre d'Arvor by 17 years, Pujadas has a faster, more expressive manner of speech. For this event, at least, public television presented news in a more upbeat and accessible fashion than the commercial station. Pujadas' news presentation from Manhattan included a somewhat slapstick encounter with an American voting machine, deviating from the conventional behind-the-desk image. Pujadas commenced his on-the-spot coverage with the following proclamation:

> Good evening Carole, good evening to everyone. For three days we will be live here from the United States under the autumn sun of New York because this election is exceptional on at least two counts. First, this is the most important election in years, or the one in which the stakes (*enjeux*) are the most serious: there are two visions of the world, two visions of America confronting each other. Second, because several hours from the opening of the polling places no one can say who will win this election; no one ventures a prediction.
>
> (France 2, Nov. 1)

The closeness of the race and the deep divisions in the electorate were dominant themes in Pujadas' prologue. In speaking of "stakes," the idea of the United States' global economic, political and military impact was suggested, similar to Poivre d'Arvor's mention of Afghanistan and Iraq. The words, devotion of time, and physical placement of news anchors in the US all implied that the election posed risks for the entire planet.

The motifs in these two prologues are most easily summarized as: *Divided America*, *Tough Campaign*, *Close Race*, and *Global Impact*. Throughout the French television coverage these motifs were related to a set of other interrelated motifs. Major motifs, such as *Tough Campaign*, were constituted through some very important sub-motifs, such as *Close Race*, and some less important sub-motifs such as *Staged Event*. Thus to distinguish between general and specific motifs we can employ a two-column table and to indicate the importance of a particular motif we can use italics (table 6.2). The motifs *not* in italics were generally tacit, lying dormant or implied by other motifs, yet implied by a positioned reading of the television coverage.

These motifs were treated in a range of tones from ironic bemusement (associated, for example, with the quaint and archaic US voting machines) to gravity (associated with the election's foreign impact) to detached curiosity (associated with the peculiarities of American culture). The motif of the candidates was a "self-evident" theme in the sense that the election appeared to be entirely about the two candidates, yet it was also incoherent and did not actually appear as a motif, that is, as a means of

Table 6.2 Major and minor themes of television coverage of the 2004 presidential election campaign. Motifs are marked in italics.

Major Motif	Minor Motifs
Divided America	two worldviews high participation rate Democrats ashamed of America and of Bush
Tough Campaign	staged event *Close Race* demanding schedule of candidate appearances no holds barred combat
Global Impact	*Democratic Deficit* *Useful Threat* *Tottering Giant*
American culture	fear of terrorism *l'Amérique profonde* "values" questions: abortion, stem cell research, gay marriage Christian fundamentalism *Continental Drift*
the candidates	*No Difference* *Big Difference*
Clumsy Voting Apparatus	incoherent voting system confusing and archaic polling machinery possible post-election dispute as in 2000

organizing discourses. Its tone was not resolved because for the French the election was not primarily about the two presidential candidates. The candidates were of interest mainly insofar as they revealed a deep rift in the electorate – that is to say they appeared in and through the *Divided America* motif. But neither man could be expected to ignore the demands for security from a fearful American populace, and the motif of *No Difference* dominated insofar as John Kerry did not appear to be planning a radical change in foreign policy. The remainder of the chapter examines the motifs in turn.

Divided America

The motif *Divided America* was evident from early September, the time of the Republican convention: "Balloons and confetti raining on the crowd at the conclusion and a touching family scene on the speaker's platform are no doubt what history will retain from this convention. But history will also retain this: to the final hour demonstrators disturbed the party" (TF1, Sep. 3). Two months later the same motif was used to help explain unusual levels of voter mobilization: "Normally not more than one American in two will vote, but this year one can say the election has impassioned America even in the most remote regions" (TF1, Nov. 1). The motif of ideological conflict surfaced throughout the period of coverage, dominating other motifs by November. We can get a sense of this motif from a TF1 reporter, Michel Floquet:

> What is involved in this campaign is much more than an economic or social program, what is involved is truly a certain idea of America. All the Democratic supporters that we have encountered have expressed an attitude of shame (I assure you the word is not too strong). They are ashamed of the image of their country. They are ashamed of this Republican administration. And when Democratic supporters encounter a Republican they shout "use your brain." [English in the original] That says it all! The cleavage is fundamental between these two Americas and it certainly explains most of the enormous, incredible mobilization that you can see at the polls across the country this morning.
>
> (TF1, Nov. 2)

The *Divided America* motif appeared in a wide range of contexts within the television news reports. It provided, for example, a neat wrap-up for a 3-minute photo essay on George Bush that aired on November 2: "For George Bush at the end of his first term: a country cut in two" (France 2, Nov. 2). It underlay descriptions of Democrats as "a bit feverish" or having "true anguish in their hearts" (France 2, Nov. 2). It permitted viewers to decipher the importance of party activists driving voters to the polls, a bizarre practice from the French perspective (France 2, Nov. 2). And it explained the "final sprint" of last-minute speeches by both candidates flying from one undecided state to another (France 2, Nov. 1). No less important was its role in directing the eyes of photographers and film editors, to serious and satirical yard signs, to door-to-door canvassers, to anti-Bush demonstrations, and above all to the long lines at the polling places (Fig. 6.4).

Halloween night lent itself particularly well to the televisual gaze, and the images all suggested a *Divided America*. Transported vicariously to the streets of New York on the night of October 31st, French viewers saw costumed revelers engaging in an odd form of political spectacle. Two masked revelers appeared: "Bush" fondling a giant globe while pulled by "Cheney" who was holding a leash. In another image, Bush sported a nose that grew Pinocchio-style, and in yet another Uncle Sam with a balloon for a head was spanking a massive "Bush." These mockeries were followed by a ghoulish looking group smeared with fake blood, chanting "FOUR MORE WARS, FOUR MORE WARS!". In the poetic words of the reporter: "While the music, the pumpkins, and the monsters are definitely present in this 2004 version [of Halloween], politics dominates the party like a firecracker. Fear for the sake of laughter is insufficient to exorcise a diffuse anguish" (TF1, Nov. 1). The immediacy and absurdity of the masquerade were uniquely suited to the televisual language and the small screen viscerally displayed the existence of a bitter political divide in the US.

For the French, disgusted by American violations of international law and embittered by America's disregard of foreign opinion, the *Divided America* motif served to prevent a simplistic targeting of disgust and bitterness toward the American population in general. If some Americans shared the French disgust then a vicarious encounter with the dissenters was conducive to a sense of shared purpose between

Figure 6.4 **American voting line scene from the 8:00 news. Source: France 2, Grand Journal 1 November 2004. Image Courtesy of France Télévisions.**

French viewers and Kerry supporters. Anti-Bush sentiment depicted in a concrete, embodied form disrupted simplistic anti-American readings of the election. At the emotional level, the *Divided America* motif encouraged French audiences to complain *with* Americans rather than about them, thus discouraging anti-Americanism.

Tough campaign

A second motif focused on the ways in which the two candidates and their respective parties had mobilized for the campaign. This motif confirmed the idea of a *Divided America* in that it framed the handling of the campaign by the two parties as a serious conflict. For the majority of the French population displeased with Bush, this motif of a pitched battle for the heart of America must have generated feelings of hope: first, the hope that Kerry might win; second, the hope that his success might produce a difference in US policy.

The image of grandiose spectacle was one facet of the *Tough Campaign* motif. This theme became clearly evident during the Republican convention, as articulated in a report by France 2's Alain de Chalvron: "Between prayers for the victims of September 11 and military posturing with combat songs and images from Top Gun, the convention definitely set a tone. The memory of the attacks was exploited to the fullest extent to bring back to life the image of George Bush with his megaphone surrounded by rescue workers" (France 2, Aug. 31). The *Tough Campaign* motif was anything but reassuring to French viewers when couched in these terms, as it indicated that Republicans were successfully playing on American fears and nationalism. The same motif underlay a description of Republican delegates as "electrified" by Arnold Schwartzenegger: "even if Governor Schwartzenegger has nothing of the traditional Republican in him. In this campaign, all forms of support are acceptable" (France 2, Sep. 1). The polish and finesse of the Republican convention supported the *Tough Campaign* motif somewhat indirectly. "With the appearance of a well-run street fair, the Republican convention reflects, in fact, a well thought-out strategy in which the feats of Arnold Schwartzenegger were crucial" (TF1, Sep. 1). Later in the election, the support of the action film actor Chuck Norris for Bush and the singer Bruce Springsteen's support of Kerry were similarly cast within this theatrical framework (France 2, Nov. 2).

Alain Chalvron employed the motif during the Republican convention to claw back the ebullience of Republican convention delegates. Following a series of short clips in which the delegates argued that Bush was popular worldwide, Chalvron declared: "There we have it, the triumph of George Bush – a triumph awaited by the convention. All the blessings of the goddesses of glory. Perhaps a bit too much. No debates on ideas. Very little discussion [*très peu de glouglouter*]. Will the Americans be convinced?" (France 2, Sep. 3). The evidence of great political tension in America suggested that such single-mindedness might not be effective.

The *Tough Campaign* motif was fleshed out with accounts of the exchanges between the two parties and their candidates. The Democratic campaign commercial featuring a veteran who was missing an arm was shown on French news with the observation that: "The time for politeness is over. Anything goes if it can help the team." Kerry's public response to the Republican convention in which he defended his military record employed "a tone he had not previously employed and that no longer spared the President" (TF1, Sep. 3). Likewise the aggressiveness of the first televised debate was noted by reporters Loick Berrou and Bruce Frankel on TF1 (Oct. 1), "The majority of American observers judged the exchanges between Bush and Kerry, particularly on Iraq, to be very tough. The two candidates let loose their aggression it seems, and Mr. Bush often appeared on the defensive." The station's preference for Kerry was only suggested at in the colorful language of the report's conclusion: "Encouraged by opinion polls, the Republicans hoped for a knock-out victory that would have assured the re-election of the President a month from now. They were surprised by the self-assurance and combativeness of John Kerry. George Bush is still the favored candidate, but his challenger may have been buried a bit too soon" (TF1, Oct. 1). The tone betrays a certain hopefulness, although it treads carefully. France 2 was less restrained, declaring John Kerry the winner of the debate but worrying that Kerry might not be able to transform his debate successes into an electoral victory. Kerry's reference to the sexual preference of Dick Cheney's daughter was framed in strategic terms as a sign of the "no-holds-barred" nature of the fight – "watch out for words that kill" – while Lynn Cheney's response was interpreted as a strategic, "well played" move (France 2, Oct. 13). The *Tough Campaign* motif explained the hectic schedule of campaign stops (France 2, Sep. 1, 3). Such signs suggested that both parties were in a frenzy (*frénésie*), that the campaign was, in the words of the reporter Loick Berrou, being conducted feverishly (*avec febrillité*), or even that the parties were scared silly (*affolement de craint*) (TF1, Nov. 2).

After the election, France 2's Alain Chalvron again recalled the *Tough Campaign* motif, commenting that "this has been a campaign with the blades unsheathed, particularly during these last months" (France 2, Nov. 3). For the long-term health of US-France relations this framing was an asset, as it prevented the Bush win from conveying the impression of unanimity and thereby bringing American culture and the American people within the scope of serious critique. The motif "contained" French displeasure by rendering it temporary and contingent.

Foreign impact

When it appeared that the presidential election might have to be settled a second time in the American court system, France 2 reporter Maryse Burgot framed the potential impact in dramatic terms: "It will therefore be a catastrophic scenario ending with a president designated by judges and an American people traumatized by the image that would give to the rest of the world" (France 2, Nov. 2). This analysis extracts signs of US instability, a motif previously introduced as the *Tottering Giant*. In fact, concern with foreign perceptions of the presidential election was not particularly high on the list of American worries in 2004. Republicans interpreted foreign critique as sympathy for terrorists, but they were contemptuous of it rather than worried, while Democrats avoided all but the most oblique references to foreign opinion, calculating that this subject could only weaken their position. This active avoidance of foreign opinion was observed, if briefly, in an interview with Pierre Moscovici (French Socialist Delegate to the European Parliament) who was attending the Democratic convention. According to the reporter: "Anti-Bush badge on his lapel, Pierre Moscovici is categorical: the socialist party supports Kerry but does not announce that fact from the rooftops" (France 2, Jul. 29). Moscovici himself encapsulates the dynamic: "what you see here, a bit, is the French being modest and discrete because our support, which is real, could be used against Kerry, and we don't wish for that to happen" (France 2, Jul. 29). The nationalist fervor that dominated American society in 2004 rendered both foreign support *and* foreign critique off limits in American debate.

Nonetheless, the discrete presence of French observers in the US was newsworthy for French audiences, and it provided an opportunity to reflect on the foreign impact of the election. This reflection often took the form of experts and authorities offering "sound bites" – very brief comments culled from longer interviews. Sound bites epitomize television's power to construct reality while appearing to offer a transparent window on the world. Both channels interviewed French politicians who had flown to the US to attend the Democratic convention as silent observers, but the two channels painted rather different pictures of those flies on the wall. France 2 presented a more politicized situation. The revealing statement from Moscovici, quoted above, indicated that French politicians were rooting for Kerry even if they must keep a low profile because of the fever pitch of American nationalism. In another interview, the author and politician Axel Poniatowski (see Chapter 4) painted the American political landscape, revealing the proximity of the Democrats to his moderate right party, the UMP: "The Democratic Party shares some things rather close to the UMP, because the Democrats are basically the party of the center-right, like the UMP, while the Republican Party is farther to the right" (France 2, Jul. 29). This explanation indicated a generally overlooked political affinity between the French Right and the American Left.

TF1 presented the same political observers as neutral and detached, demonstrating strikingly the difference created by different reporters and editors working with similar materials. On TF1, a clip of Poniatowski was selected in which he claimed, "we do not come in order to take a position, but we come to observe and make contacts," while Pierre Moscovici weighed in with the "anyone-but-Bush" sentiment, rather than the open support for Kerry he expressed on France 2: "This Bush administration does so much damage that I believe we must become involved in this necessary change – necessary for the US and necessary for the world" (TF1, Jul. 27). The only French participant to be presented as strongly pro-Kerry on TF1 was Brice Lalonde, John Kerry's first cousin and the Mayor of Saint-Briac, in Brittany, who said: "Here is someone who is not a demagogue, a guy who considers things. I've always known him to be very serious and thoughtful – before making a choice he reflects and consults. Once he has made a decision he sticks with it to the end." (TF1, Jul. 27). Through selected bits of interviews, TF1 viewers were offered three interpretations of the French presence at the convention ranging from neutral to pro-Kerry, while

France 2 with its interview selections revealed the pro-Kerry position as dominant, contextualized this in terms of the need for French observers to keep a low profile, and reminded viewers of the complicated relationship between the American and French political spectrums.

By selecting and discarding bits of interviews, public television in effect placed France in a position with political implications while the commercial channel presented an image of political neutrality that was less likely to offend. The presence of French politicians at the Democratic convention strongly underlined the importance of the election for non-Americans (albeit with significant differences between the two channels). Neither channel indicated if the French politicians were also present at the Republican convention.

Other opportunities to reflect on the impact of the election came in brief interviews conducted by both channels with intellectuals of statecraft and with authors. On October 16, Nicole Bacharan, a professor of political science spoke on France 2 of the "worldwide frustration of not being able to vote in this election despite the fact that America is the world's greatest power, and [the fact that] decisions of the American head of state have an influence on the life of everyone around the world" (*Face à l'image*, France 2, Oct. 16). This comment was extremely revealing. It indicated the tension between the subject position of the French citizen and the democratic deficit attributed to the hyperpower while suggesting France's lack of power to control certain political forces shaping domestic and foreign affairs (see chapter 1). Again, public television navigated into rather than away from controversy.

Not all interviewees were chosen to be sympathetic to the French position. France 2 had the audacity to invite Kenneth Timmerman, author of *The French Betrayal of America*, to their New York studio. Perhaps Timmerman appeared on the condition that he be introduced simply as "a journalist" rather than as the author of a virulently anti-France polemic, because his book was not mentioned on the air. News anchor, David Pujadas, interviewed him about the significance of the election and Timmerman informed Pujadas that Bush had been re-elected as a Commander in Chief "who will continue to fight against the terrorists who have attacked *us*!" When asked by Pujadas if the military situation was really the determining factor in the election Timmerman agreed, and an interesting exchange followed:

> Pujadas: "America at war?"
> Timmerman: "Absolutely!"
> Pujadas: "America in danger?"
> Timmerman: "Yes."
> Pujadas: "A fearful America?"
> Timmerman: "America is not afraid. America will pursue this conflict against those who have attacked us."
> Pujadas: "There you are then, Bush played very much on fear." [*Donc, Bush a beaucoup joué sur la peur.*]

What is interesting here is the way the viewpoint of the interviewee was casually "clawed back" to French common sense about America, particularly the *Tottering Giant* motif. Television is more likely to juxtapose than to critique, to show rather than argue, conceal rather than disprove – precisely the options employed on the fly by Pujadas against an American nationalist.

Another interview with intriguing implications featured Bernard Henri Lévy, a novelist, philosopher, and prominent intellectual of statecraft, again on France 2, November 3. Lévy insinuated that the Chirac government was merely feigning disappointment regarding Bush's victory because it "would have been very annoying for certain of them [government ministers] if John Kerry had won today," and a moment later he framed his comment in a metaphor that translates without difficulty: "therefore, watch out for crocodile tears!" (France 2, Nov. 3). This comment must be placed in context: that France had taken a definitive stance against the war in Iraq and Kerry's plan to hold international discussions could lead to demands on French leaders to cooperate with the war, which would potentially create an awkward bind for the French between domestic opinion and foreign policy obligations. Meanwhile on the European front, the project of ratifying the constitution for the European Union appeared likely in fall 2004 as long as there was an intransigent and belligerent America facing off across the Atlantic. Parts of this logic turned out to be unfounded, in large part because the French Left was much less excited about the European constitution than were the elected leaders on the Right or Left. But Kerry's internationalism, while commendable in the abstract, might take the wind out of the sails of the European project and conversely Bush was a *Useful Threat* because he could be counted on to lend credence as well as urgency to the *Europe-puissance* project.

Signs of American culture

Television news framed the election as a sign of America's cultural complexity. Not only was the US a deeply divided country, but the rupture seemed to touch on broader questions than terrorism and security, including divergent "values" regarding abortion, stem cell research, and gay marriage, among other issues. The association of religious fundamentalism with the President's position was intriguing to the French because it recalled long-standing critiques such as Talleyrand's joke about "thirty-two religions and a single meal" (Roger 2002, 70) yet it presented a paradox because of the link between secularism and democracy in the French political imaginary. The search for clues to the peculiar worldview that reconciled Christianity with militarism, private weapon ownership, and crass materialism took French reporters deep into rural America, across the South, to small towns, declining industrial cities, and libertarian refuges of the northern Rockies. Was the US about to take a turn towards religious fundamentalism and anti-modernism or was the reigning philosophy more accurately attributed to a geographically and culturally peculiar part of the US? Answers to this question found expression in several motifs.

Divided America While certain aspects of French television coverage, like the sequence of stories described in the "teaser" at the beginning of this chapter, passed moral judgments on the entire US, most aspects developed an awareness of the US as an internally differentiated space, a collection of particular places, a mosaic of political cultures. Television, because of its unique power to construct a sense of place, contributed more than other media to French awareness of internal variations in the US during the election, fleshing out regional images introduced by French newspapers (see chapter 5).

The sensitivity of Americans to regional political variations within foreign countries is weak, at best. For example, few Americans could correctly identify conservative and progressive areas on a map of France. So the French attention to regional variation of American political culture deserves attention as a highly significant *asymmetrical* feature of international communication. French television was able to construct this dissecting gaze *into* America through the combination of a place-oriented, ground-level views and cartographic, bird's eye views, as evident in the text of a November 3 report on "Bush's America."

> The electoral map clearly shows two Americas: the liberal America of the coasts and large cities and the conservative, rural America of the middle of the country. These solidly Republican states are labeled the "flying over states" [English in original] – the states one flies over by plane without ever stopping. First off there is the "Middle West," a long column from Montana to Texas. There George Bush obtained his best scores: all over 60 percent, with 69 percent in Wyoming and 71 percent in Utah. This is inland America, the home of pioneers attached to their personal freedoms, to their firearms, and to their religion… religion in fact also characterizes the traditional southern states. From Carolina to Missouri, close to 60 percent support Bush. These states where evangelists get standing-room-only crowds in mega-churches are very sensitive to questions of morality and detest east-coast liberals like John Kerry. Here the appeal of George Bush is his simplicity and proximity to the American people.
>
> (TF1, Nov. 3)

Thus *Divided America* was rendered in geographical terms, with some odd distortions of terminology and geography: "flying-over states" instead of flyover states, "Middle West" instead of Midwest, and the inclusion of the high plains and Rocky Mountains in the "Middle West." Accompanying this report was a series of images evocative of place – stereotypical but nonetheless conducive to a certain understanding of American political culture: a combine moving across a field, a small-town restaurant-café with baseball cap-wearing customers, followed by a map of the US showing red states. Next appeared Victorian houses with an antique car in the foreground, ranchers on horseback rounding up cows, heavy women with flamboyant hairdos holding hands and praying, a line of pale people praying, and again the map of the red states. Finally, George Bush in a military convocation, praying, and on his ranch. Even stripped of words these images make a clear statement: Bush country is rural, small-town America, an unsophisticated place with links to the legendary American past of the cowboy, fundamentalist sects, and the crude manners critiqued by European visitors since the 18th century.

Geographical television content most often took the form of video portraits of places and their political cultures drawn from across the United States. The small towns of Ohio, Texas, Florida and Tennessee, in particular, were scrutinized for clues to the mood of American voters and this particular election's significance. There were also scattered images from big cities, particularly New York, Boston, and Washington DC but these were outnumbered several times over by rural and small-town settings. As for the inhabitants, repeatedly the camera captured overweight people in t-shirts wearing baseball caps, straw hats, cowboy hats. Their behavior was simple: mainly eating and praying. Their primary habitats were the diner and the church – equally stark and unadorned except for wallpaper or cheap paneling (fig. 6.5).

What did these places look like to French audiences? Place experience is shaped by context, and the French landscape is replete with centuries-old brick and stone buildings. Common architectural details include heavy wooden shutters on the outside of houses, tile roofs, broad sidewalks, and narrow tree-lined streets. This

Figure 6.5 Café interior from a special report by Pascal Golomer and Tristan Lebraz in Medina, Ohio, on the 8:00 news.
Source: France 2, Grand Journal 4 November 2004. Image Courtesy of France Télévisions.

is a landscape of diverse textures, organic lines, and worn but durable materials, a landscape that visibly and often beautifully exhibits the marks of time. The low density American landscapes with their flat surfaces, right angles, and light-weight materials, broad lawns, huge parking lots, and combination of cheapness and newness contrast strikingly with the French landscape. American vehicles are different, as well; large trucks, vans, and SUVs are rare in France while France's most popular cars, a fleet of diminutive Citroens, Peugeots and Renaults, are virtually absent from the American landscape. Thus a simple camera shot of a truck in a driveway with a house in the background can convey a host of "American" associations to a French viewer.

But the televisual language went beyond conveying Americanness in general. It sought to distinguish at least two types of American place, corresponding to, and fleshing out, the *Divided America* motif discussed above. Bernard Henri Lévy offered the following portrait of urban America's response to the outcome of the election.

> It's more than a defeat. In Boston, in New York, a sort of chaos is beginning. A people, this America here, is in a state of commotion. It had such a battle and it awaited possible victory. It's Bérézina! [a reference to the humiliating rout of Napoleon's troops in Belarus in 1812] Yesterday evening, it was terrible. I saw people crying; I saw people breaking down; I saw the supporters of Kerry liquefying. John Kerry himself who was awaited to give a speech did not show. He sent John Edwards who delivered a few absurd book-keeping phrases. An impression of great sadness.
>
> (Bernard Henri Lévy, France 2, Nov. 3).

David Pujadas asked: "But there are surely at least as many people or even more who were happy?" and Lévy replied: "On the other side, certainly! If I had been in Austin, Texas there would probably have been that feeling. Because more than ever there is the effect of two Americas opposed to each other. And one of them quite clearly won out over the other." (Bernard Henri Lévy, France 2, Nov. 3) The reference to Austin (a liberal city that actually favored Kerry) is an understandable mistake. The Texas capital was meant to symbolize the heart of Bush America: interior versus coastal, rural versus urban, South and Southwest versus North and Pacific Coast. Judged by its effect rather than its accuracy, the erroneous characterization of Austin helped work against the political bias that would reduce American attitudes in general to the specific attitudes of the Bush administration.

l'Amérique Profonde The term which served as the banner for this geographical insight was *l'Amérique profonde*, a term that translates literally as "deep America," connoting rurality and remoteness. The term also carries associations of authenticity and purity, the truest representation of the national myth, the least contamination by external influences (Blyth 2005; Bernstein 1990, 21–4). But *profonde* when applied to a region also suggests a backward, reactionary & isolated society, just as "deep south" carries these connotations for some Americans (tellingly there is no "deep North"), but the French term does not indicate a contiguous region like the Deep South. "*La France profonde* as an intellectual category suggests something rather unrefined, naïve, gloomy, often religious, hokey, and immutably unmodern, the sort of place doomed (or, for the nostalgics among us, blessed) to repeat age-old agricultural patterns of life, give or take a labor-saving device here and there" (Bernstein 1990, 23). This is a term of mild derision but, in the French context, more than a little indulgence. Parisians refer to *la France profonde* "with mockery, embarrassment, and longing, all mixed together" (Bernstein 1990, 23). The key to this attitude is an urbanite's view of peasant life which, as indicated by Tocqueville in the epigraph to this chapter, appears to present the "ignorance of primitive times, rural simplicity, and rustic villages" (1969, 303). And French television coverage of the United States of 2004 found (contrary to Tocqueville) an American version of the rustic, simple and ignorant peasant in the remote corners of rural America and it was the territory of Bush voters.

The term *profonde* was applied to parts of the US in television reports as early as July, appearing surprisingly in association with Democrats rather than Republicans. A TF1 report asserted that: "In a multiracial America, Theresa Heinz Kerry, herself of Portuguese background, has easily captured the heart of minorities. But in *l'Amérique profonde*, among the most traditional currents of the Democratic party, this emancipated spouse causes some shudders" (TF1, July 28). One day later, a France 2 report used a very similar term to situate John Edwards both geographically and strategically: "Angel face and ferocious smile, John Edwards wants to seduce *l'Amérique au fond* which he comes from and where he made his fortune as a lawyer" (France 2, July 29). The term was used to situate anti-France sentiment in the United States during the build-up to the Iraq war "French wines dumped into the gutters of *l'Amérique profonde*" (TF1, Nov 3). Although the earliest uses of the term in the campaign coverage refer to heartland Democrats the term rapidly acquired an association with the Republican Party. The motif could be found in the attention given to rural locations in Texas, Tennessee, Florida, Ohio, and Iowa in reports of the lead-up to the election as well as the Election Day and its aftermath. Small conservative towns were predominant in news coverage whenever the purpose was to convey a sense of *l'Amérique Profonde*, and it was precisely the pockets of American conservatism that were of greatest interest to the French television for their exotic qualities.

But in rendering American small towns as exotic places French television also positioned its viewers closer to the rest of America – the urban, coastal, northeastern parts of the US. The coverage from major American cities was relatively scarce with the exception of Boston (which entered as John Kerry's home base). Both TF1 and France 2 posted their overseas news anchors in New York city, but the city was little more than a backdrop for the reporting of statistics and general observations on the election. Coverage of specific details of the election, such as vote hunters, yard signs, party enthusiasts, picnics, mom 'n pop restaurants, churches, and homes, was nearly always drawn from small-town America in the South and Midwest. The gaze into *l'Amérique profonde* was not usually scornful but often slightly bemused, with a sense of irony that drew the viewer's gaze to scenes most peculiar from a French point of view, such as a view of Medina, Ohio showing single family homes, lawns, and a pickup truck in the background and a yard sign in the foreground bearing the message "Elect Jesus as Your Lord" (fig. 6.6).

In keeping with this use of a generic place image to represent conservatism, George W. Bush was usually represented in small towns and on his ranch while Kerry remained essentially placeless as the news showed him standing, waving and talking in crowds. On November 2, TF1 showed Bush voting in Crawford, along with a collection of images meant to capture the essence of Crawford as an emblem of small-town America. When Kerry voted in Boston, rather than supplying a detailed place image of Boston (e.g. the Fenway, the Back Bay, the Theater District, Chinatown, the North End) the announcement was accompanied by placeless images of long lines of voters inside public buildings. Kerry's home in Boston was shown as was Boston Common but these images were brief and close range and failed to convey a sense of place. The urban, coastal America, the Northeast and the upper Midwest that supported Kerry were evidently of less interest to the French television crews than Bush's rural heartland. In any case, it was American neo-conservatism above all that demanded an explanation, and in television's language this "explanation" took the form of the *Amérique Profonde* image – which not only positioned the French

Figure 6.6 "Elect Jesus as Your Lord" yard sign from a special report by Pascal Golomer and Tristan Lebraz in Medina, Ohio, on the 8:00 news. Note the attention to diagnostic elements of the small town landscape.
Source: France 2, Grand Journal 4 November 2004. Image Courtesy of France Télévisions.

television viewer vicariously in the US, but also staked out a set of spaces marked by isolation and absence of cosmopolitanism.

When George Bush voted, French journalists were on the scene in Crawford, Texas, near his ranch, to capture the event, complete with a clean-scrubbed crowd singing "God bless America" in a parking lot. The confluence of religion and patriotism was carefully recorded for French viewers. It contributed to the exoticism because of the differing status of religion in French and American society (see chapter 1). But it also gave a place to the strange fusion of religion and politics. As TF1 reporter Bernard Volker comments:

> This small town of Crawford is sort of the reflection of America because it is a quite divided place. The mayor of Crawford and the main newspaper have in fact taken a scandalous pro-Kerry position, while four years ago they called on people to vote for their neighbor, George Bush. The consequence of this division of America is that this evening every result in a state will be contested in court if neither of the candidates clearly wins out over the other. The scenario of four years ago could very well repeat itself in many states, although certainly not here in Texas where George Bush has no worries.
>
> (TF1, Nov. 2)

French television constructed a mainstream, culturally central understanding (in French terms) of the American election. This perspective depended heavily on the image of America's interior as a place lost in the past, a refuge for 19th century values and pioneer attitudes bizarrely embedded in a slapped-together landscape of the 20th century.

Religion was clearly one of the most perplexing elements of American culture in the rural locations of *l'Amérique profonde.* On November 2, TF1 directly addressed the close ties between politics and religion with a photo essay lasting three minutes. The initial image of the essay was of a big-screen television in a "mega-church" where the theme of that week's sermon was introduced to the accompaniment of action-film style music. The audience of several thousand persons was instructed in "how to catch the ball," meaning how to obtain religious faith. The congregation prayed for the president and for the American soldiers and their families. After the service, worshippers descended by elevator to a stark basement room decorated with a pair of stuffed deer heads and a race car poster, where they sat smiling on folding chairs for a Bible study class. The entire sequence would have been completely alien to French viewers whose point of reference is the stately and traditional Catholic service wholly devoid of references to sports conducted in a Romanesque or Gothic church completely devoid of animal heads and car posters.

The film's narrator continued with an air of irony and a new image, this time outdoors: "God associated with government. There you have something bewildering in a country where church and state have been separated since the beginning. However these questions do not really disturb the Republican militants of Tennessee, reunited this day for a political barbecue. Among the guests is the pastor of the local Baptist church. He is the one who will open the meeting." The opening prayer was followed by the Star Spangled Banner which caused the overweight attendees, caught in an awkward moment as they turned in circles trying to figure out which way they were supposed to face for the anthem. Adding to the exotic impression of this scene was the metal picnic shelter, the bright red T-shirts, the plastic cups, the baseball caps – in combination these created a sense of place: ugly and graceless, flimsy and tasteless, banal and extreme, and its disoriented (in multiple ways) participants. A picnicker added to the scene by attesting to Bush's greatness on the basis that he prays every morning for an hour and uses prayer to direct his political decisions. The probing eye of French television juxtaposed disoriented Americans with misguided policy, at least from the intended subject position. The photo essay not only revealed what was alien in American political culture but it *placed* that alien culture in the American landscape. This positioning of the viewer was incompatible with simplistic anti-Americanism, however, because this was clearly an odd place, almost as different from Boston as from Paris.

France 2, in particular, devoted time to the geography of conservatism, finding its essence in various forms in Noxon, Montana where it appeared that people still wore coonskin hats, and in Kennesaw, Georgia where gun ownership was required by law. In Noxon, the camera captured a town fair complete with bizarre customs such as horse-back watermelon slicing and the cooking of gluey grits in enormous kettles for communal consumption. But the narrator neglected to mention that the observed activities were part of some kind of festival and that the participants were dressed in costume (fig. 6.7).

The aftermath of the election again brought attention to *l'Amérique profounde*, this time to explain the unsettling outcome of the election. Reporter Marine Jacquemin judged that Ohio voted against its economic interests by voting for Bush, preferring instead to support values which she summarized as religion, the death penalty, the bearing of arms, and opposition to abortion. When David Pujadas interviewed Kenneth Timmerman and asked him if the outcome of the election was not "a bit the revenge of *l'Amérique profonde*?" (France 2, Nov 3), the conservative author only provided a non-sequitur about Americans preferring Fox news to other stations. Probably the term *Amérique Profonde* was meaningless to Timmerman because Americans divide their country into regions such as South or Midwest rather than employing cultural frameworks with unbounded geographies.

The term *l'Amérique profonde* was, in short, a French geographical construct applied to the US to make sense of the political culture in places that most strongly favored George Bush, and as such it was evocative of both place and personal identity for the French. In each place, the inhabitants – their expressions, their dress and mannerisms as much as the streets and landscapes – captured the essence of deep America, a half-civilized place proud of its armed, self-sufficient, Protestant, heterosexual normality. This "normality" was instantly ironic on French television, however, because of the visual excess – as if *Sixty Minutes* depicted French nationalists wearing berets and eating escargots.

Although this was not a well-rounded view of conservative America it *located* conservatism and discouraged simplistic anti-Americanism. Not all motifs did so.

Figure 6.7 Men talking at a picnic table in Noxon, Montana. Note that the men are wearing cowboy and coonskin hats, a rifle butt is visible in the foreground, and a red, white and blue braided rug is hanging over a rope in the background. Special report by Loic de la Mornais and Emmanuel Becque, on the 8:00 news.
Source: France 2, Grand Journal 24 October 2004. Image Courtesy of France Télévisions.

One that was supportive of anti-Americanism is based on the idea of a growing rift between the US and Europe.

Continental Drift David Pujadas captured the image of cultural divergence across the Atlantic by referring to a *fossé* – that is a ditch, trench, or moat – opening between the two continents. He was not alone in envisioning a kind of *Continental Drift* widening the gap between the US and Europe. While the motif of *Divided America* encouraged sympathy with roughly half of America (Kerry supporters) the motif of *Continental Drift* in essence dismissed America's internal divisions as irrelevant to the future and therefore to geopolitics. What was winning out in the US was the America of the conservatives, *l'Amérique profonde* with its passion for guns and mega-churches. This was the future of America, and in this future the best Europe could hope for was a cordial détente rather than a close partnership.

To some extent the perceived rift between the continents was a product of deteriorating relations between the French and American governments. A TF1 report of November 3 began with the image of Bush and Chirac embracing, and the comment:

> This huddle marks a moment of grace! On the 18th of December, 1999 Jacques Chirac was the first foreign head of state to congratulate George W. Bush on his electoral victory, even before Bush had taken office. Less than two years later Chirac was the first foreign leader to visit the ruins of the World Trade Center. This was the era when we were 'all Americans.' Then the great fracture. French wines dumped into the gutters of *l'Amérique profonde*, French fries re-baptized 'freedom fries,' and the candidate Kerry caricatured at the Republican convention as a 'French poodle.' Between these two periods, the war in Iraq and the great rupture.
>
> (TF1, Nov. 3)

While television so effectively captured *l'Amérique Profonde* and its inhabitants in images, it was able to convey the idea of *Continental Drift* only through interviews. This made the latter motif less salient. Francois Bayrou (President of the center right party, UDF) declared that: "We have to really ask ourselves about who are the American people, who are fundamentally very different people than the French and European peoples" (France 2 Nov. 3). Dominique Moisi (special counsel to the French Institute of International Relations, IFRI) expressed the belief that America: "Has invested itself in a divine mission to reestablish a sort of religious and moral order in the United States. This America will have less and less in common with, and fewer relations with, Europe. We belong to two different emotional continents" (France 2, Nov. 3). David Pujadas followed up with a question to Bernard Henri Lévy: "Bernard, do you also have the sense that the America which has triumphed today is drawing farther and farther away from Europe. Could it at least be the inverse?" (France 2, Nov. 3). His question indicates the possibility that it is Europe that is drawing away from the US, rather than vice versa. Henri-Lévy replied to the question as follows:

> I also believe, in effect, like Dominique Moisi; I believe the problem will be less the war in Iraq than the interior crusade. I have listened throughout the whole campaign. It's true, I've probed this *Amérique profonde*, and the questions of values, as he says – that is, the question of homosexuals, the question of women, the question of abortion – have dominated the campaign of George Bush more than foreign policy. Thus today it's not an election it's a plebiscite, and it's a plebiscite in favor of the values of a certain America.
>
> (France 2, Nov. 3)

It is the US that is changing, we are told, and its movement is towards *L'Amérique Profonde* which has won the battle and unfortunately will win the cultural civil war. Pierre Lellouche, a spokesperson for the UMP, refered in another interview to: "A stereotype of America that we loved in the Cold War past – internationalist, patient, polite, listening to Europe" and dismissed this as obsolete with a few peremptory words: "they are no longer listening to Europe" (France 2, Nov. 3). The motif of *Continental Drift* arose only through the means of interviews and therefore was borrowed whole cloth from the communication sphere of scholarly discourse (see chapter 4). Intertextually, television drew on this motif to support other motifs more germane to televisual discourse, but its pessimism and lack of geographical specificity – its placelessness (to use a term applied somewhat differently by Relph 1976) – did not fit in with the other motifs of the television coverage, and hence it never rivaled the other motifs in terms of its prevalence. Simply stated, this was the closest French television came to anti-Americanism.

The candidates

Finally we must address a motif that would seem "naturally" to dominate coverage of an election but which in fact was neither clear nor dominant on French television – the candidates. Much as in the newspapers, what did emerge was a pair of contradictory motifs, *No Difference* and *Big Difference*, in relation to the candidates and contextualizing them in terms of the motifs previously discussed. *The Candidates* is therefore a *subsidiary* motif linked to the other motifs in various dependent ways, not a dominant framing device. With regard to the two presidential candidates we find a kind of vagueness that is not present in the other motifs. French observers made no effort to hide their disgust with the Bush administration but they were not certain whether they should place any more faith in a Kerry administration. Therefore they oscillated between noncommittal expressions of hope that Kerry might bring a change of some sort (domestic or foreign) and grim pronouncements that this difference might only involve rhetoric and therefore produce few tangible policy changes. Yet the outrage at Bush administration policies and rhetoric was strong enough to overcome this uncertainty at times. The oscillation between pessimism and optimism demonstrates the French tendency to think in shades of grey rather than black and white.

On November 1, France 2 displayed photographs of the two candidates, side by side, with their ages, family status, the years of their education and alma mater, and indications of their religious convictions. A voice-over indicated how similar the candidates were on each of these points and David Pujadas only hinted at the existence of differences. The next day, however, *Big Difference* reigned as France 2 presented video profiles of the two candidates, each lasting three minutes. The Bush profile presented Bush as having a unique presidency: "The first president designated by the Supreme Court; the first president to preside during an attack on American soil; the first president to engage in a 'preemptive' war, and the first president to provoke a virtually global sentiment of anti-Americanism" (France 2, Nov. 2). Four separate moving images accompanied this framing of the Bush presidency in the four quarters of the screen: Bush taking the oath of office, a plane hitting the World Trade Center, the toppling of the giant statue of Saddam Hussein in Baghdad, and an anti-war banner unfurling over a MacDonald's sign. The profile mentioned France's opposition to the Iraq invasion: "The world divides into two camps: the camp of peace led by France, Russia and Germany, and the camp of war led by the US, followed by Great Britain and Spain" (France 2, Nov. 2). It also mentioned the role of neoconservatives (*neoconservateurs*) in shaping Bush administration's foreign policy and it offered a body count for Iraq: "a hundred thousand civilians and a thousand American soldiers. Certainly Saddam Hussein and his dictatorial regime have fallen, but the ensuing chaos has given a bitter taste to what George Bush labels a 'military success'" (France 2, Nov. 2). This pessimistic segment was followed by film clips of marching anti-war protesters and photographs of prisoners being abused in the Abu Ghraib prison. The narrator remarked: "the behavior of some of their soldiers suggests that the US has lost all sense of human and democratic values" (France 2, Nov. 2), but the narrative ultimately clawed back this discomforting sense of an ally gone bad by imposing the *Divided America* motif and eliciting the image of "a country cut in two."

A video portrait of John Kerry followed, and helped complete the claw-back. The Democrat was described as a star pupil, a heroic soldier, a good father, and a fighter for just causes, attributes summed up by the moniker, "*monsieur trop c'est trop*," and the English translation: "Mister Too Much." If this comment seemed snide or supercilious the producers hastened to add: "It seems incredible, but it's all true" (France 2, Nov. 2). Footage of Theresa Heinz Kerry reflecting on her childhood, speaking in French, is rounded out by a description of her international background and "*franc-parler,*" her bluntness and honesty, and the insight that she helps "seduce" those who find her husband too straight-laced, complicated, or distant. Here then, are the elements of a different America: ambitious and driven like the Republicans and their leader, but morally committed, cosmopolitan, and disarmingly honest.

The candidates are given meaning in these profiles by the divided America motif. From the French point of view, Kerry's cosmopolitan qualities are the more presidential qualities whereas Bush's militarism is a clear shortcoming. Therefore Kerry's success was something to be hoped for. But what would be the result of his potential success? Immediately following the video profile of Kerry, Philip Gordon of the Brookings Institution appeared on camera with David Pujadas to offer his analysis of the two candidates, and in effect to reinstitute the ambivalence of the *No Difference* motif (France 2, Nov. 2).

> Philip Gordon: In effect, I think we would be deluded to expect a massive change of US policy in Iraq. John Kerry thinks like George Bush that it would be dangerous to withdraw the American troops. He thinks it necessary to hasten the training of Iraqi security forces, like George Bush. And he thinks elections [in Iraq] should be organized for early next year, like George Bush.
> David Pujadas: No difference up to that point.
> Philippe Gordon: That much said, uh, up to that point, no difference. But I think all the same there will be a difference in perspective. George Bush thinks and says and gives the impression that everything is going well in Iraq, that he has not made mistakes, etcetera. Kerry at least – and above all with allies like France and Europe – could say 'yes, I agree with you, we made some mistakes but now we have shared interests and we must work together.'

The result of this framing is that the questions relating Bush to Kerry remain in shades of grey. Other questions, like the attitudes behind Bush administration policies, the sentiment of conservative Americans, the divided state of American society, the passion surrounding the election, and even the future of US-France relations, presented a clearer image. *Continental Drift* is subordinate and *Divided America* is dominant, conveying a sense of perspective on the outcome – whatever it might be – putting the candidates in perspective and encouraging a sense of America at war with itself and slightly deranged in the wake of its first home court loss.

Through television, French audiences could not help but acquire a sense of the radically different style and attitude of the two men running for president of the United States. At the same time America could hardly be expected to pursue a radically different course, regardless of whose hand was at the helm.

Clumsy voting apparatus

A final motif bears mention for bringing a touch of levity to the television news. This was the portrayal of the mechanical side of the American electoral process. French viewers were treated to a view of odd, anachronistic and clumsy voting machines. The motif drew intertextually on news coverage of the errors and confusion of the 2000 election. A couple who had recently immigrated to the US from France was interviewed on their experience using American voting machines, which they called a *casse tête* meaning a difficult puzzle (literally "head buster"). With a touch of slapstick, David Pujadas actually demonstrated the operation of a lever-style voting machine that had been borrowed for the broadcast.

Here clawback was in full force. While America might be powerful it was also primitive – that was the message of *Continental Drift* and also, to a degree, the message of *Divided America*, as well as aspects of the American Culture motif (especially *l'Amérique profonde*). While America was strong it was also prone to collapse, as indicated by the *Tottering Giant* motif. So the early machine age ambience of the *casse-tête* voting machine sealed up the narrative by offering another sign of primitivity. Behind the curtain the hyperpower was depending on some rather rickety machinery.

Conclusion

If it was reassuring to hope that Kerry might win, a fair amount of anxiety remained, based arguably on America's power and on the democratic deficit. This tension sought release on television in a bit of irony: although America was the world's most powerful democracy its citizens conducted odd rituals while dressed like savages or poltroons, and its democratic machinery – voting booths, electoral college, judicial involvement, and so on – were clumsy and antiquated. Nonetheless, French television coverage exploited very few of the available opportunities to present a negative view of Americans in general or the United States as a whole. Anti-Americanism probably underlay the sequence described at the beginning of the chapter, which sandwiched a report on the Cheney-Edwards debate between coverage of the war in Iraq and a special report on the death penalty. It may also have driven the asymmetry in the television news reports such that more attention was given to *l'Amérique profonde* than to the primarily democratic parts of the country, but the television coverage also emphasized what was most intriguing and therefore most "newsworthy" for French viewers. The motif of *Continental Drift* carried these tendencies to an extreme, but this negative interpretation of the US was subordinate to ambivalent constructions in terms of time and televisual appeal.

The *Divided America* motif and *Tough Campaign* motif depended on the existence of the Democrats, in effect contradicting the *Continental Drift* motif and demonstrating that if the US were in fact drifting away from Europe in cultural terms there were many Americans trying to stop that drift. The *American Culture* motifs, while focusing on *l'Amérique profonde*, showed that the most alien and disturbing aspects of America were rooted in particular places rather than endemic; the roots of Bush's unilateral foreign policy were not found everywhere in America, at least not to the same degree. The problem lay in the most isolated parts of America.

Television's ability to portray people and places and its limited affordances with regard to other, less personal knowledges, created a particular kind of ambivalence about the election. The result, while emphasizing the most conservative places and lifestyles in the US, also showed the provincial, violent and messianic aspects of American culture to be place-bound and contingent. The frightening images of righteous piety and sanctimonious aggression displayed to France and the rest of the world by Condoleezza Rice, Donald Rumsfeld, Dick Cheney and George Bush were not emblematic of all of America but only of *l'Amérique profonde*. Furthermore, this deep America was locked in a dead heat with other parts of America. These other places – the coasts and the Democratic strongholds of the northeast and upper Midwest – may have been much less visible on French television in fall 2004, but their existence was constantly recalled by the *Divided America* motif.

The *Divided America*, *Tough Campaign*, *Foreign Impact*, and *American Culture* motifs coexisted in a certain tension or ambiguity. This ambiguity, in turn, drove the representation of *The Candidates*. On the one hand, John Kerry was clearly more *sympathique* (like-minded), but on the other hand a *sympathique* American president might not actually be a blessing for France's European objectives or for global democracy. Unless America was going to undergo radical changes, such as a fundamental and lasting renunciation of the use of economic and military powers, Kerry would simply put a congenial face on the *hyperpuissance* and this would sap France's strength to develop a new center of world power based in Europe. Although this worry was not articulated in television news, it remained behind the surface of the screen as a kind of force field organizing television's normalizing discourse of sounds and images. The potential that the election, for all its drama, might not present a real difference from the French point of view in turn may have lent some attractiveness to the idea of the US's vulnerability as framed in the *Tottering Giant* motif.

Differences between public and commercial television were evident (although less pronounced than in the US). Public television was more willing to take a position with political implications while commercial television took the safe road of "neutrality" as demonstrated in the selection of quotes from French politicians present at the Democratic Convention. Public television also undertook a more sustained examination of the election and of American society, with more news coverage and more serious offerings outside of the news, while commercial television appeared to devote a similar amount of time to the election but did so mainly by means of its all-night coverage of the election that was not particularly informative. Public television revealed its greater capacity to make up for the democratic deficit. Insofar as democracy is eroded by transborder economic flows and the "hollowing out" of the state to serve the global market this finding is not surprising. The economic context of public television allows it more freedom to play a real political role on the side of democracy and against forces of economic globalization.

Chapter 7

Internet: Ideal Speech Situation or Babble?

> In a democracy an association cannot be powerful unless it is numerous. Those composing it must therefore be spread over a wide area, and each of them is anchored to the place in which he lives by the modesty of his fortune and a crowd of small necessary cares. They need some means of talking every day without seeing one another and of acting together without meeting.
>
> de Tocqueville, *Democracy in America*, p.518

Although the quote above was written in reference to the newspaper, the Internet clearly fulfills the purpose of "talking every day without seeing one another" more adequately than the newspaper. Through many applications such as e-mail, discussion lists, and chat rooms, the Internet functions as a political forum – a space in which conflicting ideas about policy, governance, ideology and personal identity are discussed and debated. The Internet permits what we could call "scattered gatherings" of citizens, creating the possibility of reconstituting the public sphere among spatially dispersed communicators (Dahlberg 2001; Froomkin 2003; Kellner 2000; Poster 2001; 1997). During the 2004 election, French citizens engaged with each other and with the US election in several online virtual spaces, advancing alternative constructions of the United States, American society and American leadership. These representations of America as Other, whether as a leading force for global democracy or as the *hyperpuissance*, offered perspectives on French society, culture, policies and leadership, as well.

Like a physical gathering place, the Internet brought together observers of the 2004 presidential election in a social situation with a collectively recognized purpose – the interpretation of a political event. Also like a physical gathering place, the Internet shaped the interactions between its "occupants" in certain ways, lending some participants a kind of authority while affording little authority to others. The Internet therefore acted as a kind of *social structure*, shaping the social interactions occurring within it by enabling certain kinds of interactions and discouraging others (Sclove 1995). The "occupants" of the online debate were able to present themselves in certain ways, taking on particular roles and identities in ephemeral communities that formed around brief (but often heated) debates. In short, it hosted scattered, fluid, anonymous, free-form political gatherings. A range of such virtual gathering places on the Internet facilitated the formation of a kind of public sphere (Habermas

1989) where groups of participants worked toward fleeting and contingent forms of consensus.

The key to understanding this communication is that French citizens encountering a loss of power due to an erosion of French sovereignty under capitalist globalization could regain a sense of control in the face of distant and potentially quite important political processes through participation in the *symbolic framing* of the election (Gitlin 1980). Hénaff and Strong (2001, 222) characterize political involvement in the "virtual space" of the Internet, noting that its function is not to make the real virtual, but quite the opposite, to make the typically virtual public debate real. On the Internet: "No one is simply in a passive receptor position. One can answer, question, approve, and thus debate. It is a space given over to agonistic exchange."

The Internet held a special place within the panoply of media appropriated by French citizens to discuss and debate the election of 2004. What stands out in the Internet context, relative to other media representing the American electoral drama, is the frank and passionate way in which the motifs of American power and French vulnerability were articulated. These expressions were indications of a larger process of identity construction through a range of communication media. Discussions on the Internet symbolically constructed the American election, society, culture, and leadership, but they also constructed the French observer as both outside (a critical, and often cynical observer) and inside (a subject of the American empire chafing for the rights of citizenry). The latter construction was substantially more prevalent on the Internet than on the other media, although nothing like consensus emerged from the Internet. Quite the opposite, the virtual space of Internet appeared to encourage acrimonious debate insofar as the people with the least popular points of view were often the most "visible" participants. Compared to television and newspapers, Internet discussions often represented America as an entity while seldom invoking the idea of America as a collection of people differentiated by place and personality (as seen previously, for example, in the *l'Amérique Profonde* motif). The threat posed by the United States to French sovereignty and the concept of democratic deficit were, not surprisingly, most apparent in this medium.

The volume of written material generated on Internet discussion forums relative to the US election is rather impressive. On *Le Monde*'s forum there were over 1,000 separate discussion threads and over 10,000 individual comments or "postings." On Wanadoo (the French equivalent of America Online) there were 11,890 postings. The forums sponsored by *Libération* and *Le Figaro* together added another 4,000 postings, to raise the combined total to over 25,000 postings related to the US election. To derive a sense of the nature of this international, multilingual discussion about the US on the eve of the 2004 election I have employed both qualitative and quantitative analyses.

For the quantitative portion of the study I adopted a sampling methodology to determine the balance of general preferences (pro- and anti-) in these discussions regarding key issues: the candidates, the US, France and Europe. Although topics of France and Europe entered the discussions tangentially (the forums under consideration being ostensibly about the US election rather than France or Europe) they were consistently evoked by the discussion of the US presidential campaign. I compared several discussion forums and aggregated the data to determine both the relative percentages of various positions and the degree of preference expressed on each of four themes (clusters of motifs).

For the qualitative portion of the study I analyzed discussion forums and blogs and translated segments of discussions or entire discussions. In translating these discussions I looked for representational motifs, as in the earlier chapters. These could not be identified with the same level of confidence as in the cases of television and newspapers because on the Internet the range of topics was significantly broader. With only a bit of flippancy one could characterize Internet discussions as suffering from "attention deficit." It was not out of the ordinary to find a comment about trade followed by a complaint about militarization, and that in turn followed by an observation about politics, each of these comments linked only weakly to the others. Because of this attention deficit, a theme that appears in one posting, even the original posting in a thread, is unlikely to reappear in more than perhaps 1/10 of the subsequent postings. Therefore the motifs I identify in this segment are not *structuring* motifs as much as *connecting* themes. They string together weakly related topics in a single Internet forum and they connect any given forum to motifs in other media such as television, the newspaper, and scholarly writing. Despite this difference, many of the connecting themes will recall the motifs in previous chapters.

Affordances of the Internet

What began as a form of insurance against the possible disruption of the US military command system by a nuclear attack quickly, and with a certain irony, became a virtual space for ad-hoc community building (Rheingold 1993; Adams and Warf 1997). From its earliest manifestation as ARPANET, computer networking was devised so as to permit intercommunication without a single dominant relay center or hub. In addition, network architects attempted, and largely succeeded, in minimizing to the point of imperceptibility delays based on distance although delays based on other factors such as bandwidth and server speed continue to be perceptible. Still, most such delays are measured in seconds rather than minutes, rendering the Internet the first viable long-distance, multi-user, real-time discussion forum. To a significant degree, then, the Internet generates an opportunity for factors other than location to structure social interaction. If instantaneity and the annihilation of distance permit new forms of gathering, the intervention of the keyboard in many Internet applications (such as chat rooms, discussion forums and "blogs") further transforms the nature of gathering. To gather in a shared communication situation no longer entails being aware of the physical bodies or real identities of the other people "present," although imagination tends to fill in this gap in knowledge (Turkle 1995).

Thinking of the Internet as a context for political communication means we can envision something that allows people to gather, and in this sense is like a physical environment such as a meeting room, lecture hall, or town square, but that does so in a way that is short-lived, constantly under construction, and dependent on anonymous, scattered, and "distanciated" (Giddens 1984) participants. I have argued elsewhere that the Internet provides "new contexts in which people take action: virtual places of many sorts" and have demonstrated the topological similarities between various Internet communication protocols and a range of physical places (Adams 1998). Instantaneity and anonymity are combined with various communication topologies, that is, particular arrangements of nodes (communicators) and links (communications) (Adams 1998). "The Internet" is in fact a family of topologies: one-to-one communications (e-mail), one to many communications (blogs and other personal websites, and mass e-mailing or "spam"), and many to many communications (discussion forums and chat rooms). In this regard, the Internet mimics telephone, radio, newspaper, and a wide range of physical places from the private room to the meeting hall and town square (Mitchell 1995).

As ARPANET evolved into the Internet between the 1960s and the 1990s, this new virtual space knit together communities based on political views, sexual preference, environmental interests, religious conviction, and ethnicity among other things (Warf and Grimes 1997; O'Lear 1997; Starrs 1997; Adams and Ghose 2003). Although such communities lack certain features found in place-based communities they have all of the core characteristics of community such as on-going communication, sharing of resources, sharing of interests, interdependency, establishment of norms, and a sense of belonging (Wellman and Gulia 1999). It is easy to condemn the absence of community in online social interactions but doing so implies an idealized pre-Internet community that has rarely if ever existed. In both physical and virtual communities there are problems with power differentials, exclusion, manipulation, deception, and the like. Neither type of community assumes perfect agreement (despite claims to the contrary, e.g. Young 1986) but each of them in its own way is a context for seeking consensus – an unending endeavor – and therefore a part of what Habermas calls the "public sphere" (1991).

The Internet, as a global system, has been blamed for forcing communications into a globally homogeneous form, promoting a loss of place-specific experiences. It would be mistaken, however, to make this connection between the scale of the communication infrastructure and the scale of the identities people construct in and through that infrastructure. Hénaff and Strong frame this point cogently: "The network model that prevails in connected systems permits a previously impossible affirmation and acknowledgement of the local" (2001, 227). They follow with the argument that the Internet is a new type of public space not because it is virtual, but rather because it "permits and integration of individual spaces into a common one without thereby emptying them of their singularity," which opens nothing less than "a new chapter for democracy" (Hénaff and Strong 2001, 227).

Social interaction on the Internet is somewhat peculiar, if not entirely unique, because in physical contexts of interaction distance translates into time, money, or labor that must be expended to maintain communication. Once one is "in" the social context of the Internet these distance-related communication costs are minimized, although the initial access costs of a computer and Internet access remain. Once one has crossed the barriers to Internet access, it is relatively easy to cross between radically different kinds of online "places" and enter a wide range of discussions with little personal cost in time, money, risk or inconvenience. The power of proximity as the primary force shaping political culture gives way to network connectivity. Shortcuts between "distant" sectors of the Internet with opposing interests and viewpoints are greatly outnumbered by links between "nearby" sites that cater to people with similar worldviews. This connectivity pattern is both cause and consequence of a sorting out of users according to political stances and belief systems and the production of a new sense of place. On sites that support debate, such as online forums and blogs, persons generally share more or less the same worldview. Thus it is possible to speak of distinct "zones" of political debate on the Internet as well as intertextual links or bridges between the Internet spaces, and intertextual links to virtual spaces in other media. Within a zone of essentially like-minded people, however, the costs of dissent are low, so it is not uncommon to find dissenters and angry but ineffective attempts to put them in their place with "flaming" (online insults and aggressive verbiage).

Integral to the constitution of new identities via the Internet is the restructuring of community. Whereas previously the term "community" implied propinquity, it depends increasingly on shared interests and/or participation in on-going communication situations. Howard Rheingold (1993) drew attention to this phenomenon over a decade ago, with his study of the various forms of community life that sprang up in the WELL (Whole Earth 'Lectronic Link) including former hippies, Deadheads, and veterans of communes, as well as the technologically gifted. Such users interlinked to form new communities, and new *types* of community. In some cases the Internet was also used to enhance citizen access to a place-based community, such as the Blacksburg Electronic Village, one of the oldest online networks designed to enhance civic access and public participation in the town of Blacksburg, Virginia. But the term "community" does not capture all of the unique characteristics of online communication. For this we may need new concepts and new terms, such as the idea of "bridgespace" (Adams and Ghose 2003) a term coined to capture the Internet's bridging function between two or more places. A bridgespace is a "variegated space of international and multicultural communication… a set of connections between here and there, in both a geographical and a cultural sense, like a rail yard or an airport" (Adams and Ghose 2003, 415–16). What is indicated is not simply a "space of flows" in Castells' terminology (1996) but rather a particular link between two places (like France and the US) and an associated way of engaging in the world. Just as one can ground one's sense of identity and one's moral commitment to a particular place, one can also ground identity and commitment to a "space between" (Adams and Ghose 2003) that is in effect a combination of a mediascape, ideoscape, financescape and technoscape – to employ four of the globalizing contexts of perception and action identified by Appadurai (1996). A certain kind of bridgespace is formed when citizens of one country vicariously participate in the events of a distant election, enhancing reverberation between polities and supporting political transnationalism. Such transnationalism consists of the online sharing of links to web pages, quotes, and images from the distant country (in this case the US) as well as discussion and debate about the meaning of this symbolic material.

The Language of Internet

The language of Internet communication has various peculiarities with political implications. Many Internet service providers charge by the minute, so users develop methods for minimizing the time it takes to convey an idea. People typing comments into Internet forums, for example, are inclined to abbreviate common words, employ incomplete sentences, ignore typographic errors, and so on (Lanham 1993). Furthermore, many Internet-based communications are meant to be read once, not repeatedly as is the case with material in print. People are inclined to frame ephemeral bits of writing in a conversational style closer to speech than ordinary writing. Table 7.1 provides a sample of some abbreviations used in French Internet discussions. This list is not exhaustive but it gives a sense of the prevalence of this phenomenon which is also quite common when the Internet is used to communicate in English.

Table 7.1 Examples of abbreviations used to write entries in French on the Internet, collected by the author from various discussion forums and blogs. In some cases this system benefits from a homonym; the letter "c" is pronounced "say" in French, and the contraction "c'est" (it is) is also pronounced "say," so the letter can easily stand in for the contraction. In most other cases an abbreviation is created by dropping some or all of the vowels from a word.

Internet abbreviation	French word	English equivalent
ricains	Américains	Americans
bcp	beaucoup	much/many
c	c'est	it is
cad	c'est à dire	that is to say
ds	dans	in
il fo	il faut	it is necessary
pcq	parceque	because
pko	pourquoi	why
qqchose	quelque chose	something
qqes	quelques	some
tjs	toujours	always
tt	tout	all
tte	toute	all
ttes	toutes	all

The ephemeral quality of Internet writing inclines people to make other alterations in the signifier-signified relationship, as well, for example using a temporary screen name (an assumed alias) that contains a clue to the meanings they want to convey at the moment rather than a strong permanent link to the self like one's ordinary name. In other words, I could join one discussion as "Disgruntled Atheist" and another as "better red than dead" to chat about religion and politics, respectively.

The use of abbreviations may introduce a certain amount of indeterminacy among those readers who have not previously encountered a particular abbreviation. Even more likely to generate confusion is the tendency to truncate sentences and thoughts and the rarity of proof-reading on the Internet. At times, users intend to create confusion and ambiguity in order to interject *something* in a discussion but not necessarily a clear proposition. In rapid-fire, anonymous discussions, the effects of ambiguity can be very interesting, and users may be more interested in discovering what will happen than in trying to convey a particular idea. So the ambiguity inherent in saving time and money is combined with ambiguity founded in the ephemeral quality of this writing and finally the ambiguity people create for fun or some other form of gratification. Internet discussions therefore do not contain an orderly set of motifs as encountered in previous chapters and searching for them would not be meeting Internet writing on its own terms.

But the Internet is not only a verbal medium. It frequently carries images and occasionally sounds. These are often "borrowed" from other media such as magazines, television, and music CDs. They make the Internet *intertextual* (Kristeva 1980; Barthes 1974) as they encourage cross-pollination between texts directly through links and indirectly as web-users make mental connections between content on various sites, on and off the Web. While images and music can be made to load directly onto a webpage that "borrows" these nonverbal texts they can also be linked from one page to another page creating *hypertext*. Because of intertextual and hypertext links any given "communication" on the Internet is likely to re-present or cascade into other texts. In some cases this reduces the role of the individual communicator, who does little more than provide a link to another person's text or image, and therefore minimizes the role of "author." In other cases, the combination of linked representations with original text may add to the originality and usefulness of a communication. In short, the text does not stand alone on the Internet but is profoundly intertextual. This means the author is not an "authority" in the mold of the author of a book, essay or poem.

The Internet as Public Sphere

Jürgen Habermas has proposed the conditions for an "ideal speech situation" that provides a point of reference for judging the constructiveness of online communication in developing the public sphere. These conditions are that every subject with the competence to speak and act is allowed to take part in a discourse and participants in a discourse are allowed to raise any question, make any assertion, and express any attitude, desire, or need without interference from internal or external coercion (Habermas 1990) (see chapter 1). As previously noted, the Internet permits rapid (real-time) discussion between participants who are separated from one another in space and who remain anonymous (through the use of pseudonyms). This in turn permits participants to speak without fear of being located and physically or emotionally persecuted for the views they express. It also permits participants to remain unidentified as male or female, white or nonwhite, young or old, handicapped or non-handicapped, and so on. The possibility of external coercion is thereby reduced, and social power relations originating outside any given discourse are concealed by distance and technology, which reduces the justification (if not the actual habit) of internal (self) coercion. In addition, the informality of the Internet "setting" encourages people to disregard social rules and conventions and to raise questions that would otherwise be self-censored due to notions of etiquette, although a weak version of online etiquette called "netiquette" exists (Rheingold 1993). In sum, the Internet seems a likely "space" or "place" for the construction of the public sphere.

However, if people are inclined to "speak" their mind without fear that someone will call them to order on the Internet, this lack of authority and order also creates a favorable environment for irrelevant or tangential comments which tend to diffuse the focus of the conversation and reduce participants' ability to address pertinent issues. Intertextuality on the Internet breaks down social hierarchies that can stand in the way of an ideal speech situation but it reduces the possibility of rational debate, leaving open the question whether Internet can actually support an "ideal speech situation." The political challenge for Internet communications is how to maintain *sufficient focus* so that points can be made and reflected on despite "flaming," jokes, rumors, and other manifestations of communicational chaos.

A consequence of the lack of sanctions in the Internet communication environment is what I call the *principle of the noisy outlier*. Normally when someone takes a minority position in a conversation he or she is subject to various forms of coercion by the majority. If I try to convince people in a physical gathering like a party that the earth is flat I will be mocked and ridiculed, which would happen on the Internet, as well. But in physical gatherings I will also be "told" through body language and voice tone that I am beginning to overextend my welcome; voices will be raised, jokes will be made at my expense, laughter at such jokes will indicate a coalition against me, and finally my listeners will drift away to other parts of the room in order to effectively remove the outlier from the conversation. Since one is seldom anonymous in face-to-face communications, the negative impression may last and contribute to a "bad reputation" which reduces my credibility, handicapping my communications. Communications of formal kinds, such as meetings, involve even more rigorous methods for silencing the outlier. The chair of the meeting is likely to "manage" a dissenting individual by saying "we need to move on," or forcefully re-articulating the majority opinion, or calling for a vote or even requesting that someone make a motion to adjourn. On the Internet the noisy outlier is relatively safe from silencing tactics and does not risk his or her reputation by being outspoken or even obnoxious, because it is always possible to create a new "screen name" and this name does not necessarily bear any connection to an enduring personal identity. This means that more popular perspectives and less popular perspectives tend to receive rather more equal amounts of attention than in comparable physical gatherings.

The study of Internet discourses therefore permits consideration of communications that are weakly structured, informal, and nearly ideal (in the Habermasian sense). But they tend to be raucous, disjointed, and highly uneven in their degree of reflection and constructive engagement. Noisy outliers deliberately discourage the discovery of common ground or domains of consensus. The least constructive form of communication, "flaming," is intended to intimidate other participants and works to destroy the ideal qualities of the Internet speech situation by introducing a threat of violence which, even if it cannot be converted from screen to physical contact, nonetheless results in some degree of intimidation or exclusion, deteriorating the public sphere.

In addition, the linearity and sequential flow that are the norm (albeit frequently violated) in face to face discussions (people take turns talking and respond to each others' ideas in turn) are not typical of Internet discussions. Whether in "chat rooms" or discussion forums, Internet discussions flit from topic to topic, with participants addressing whatever observations or ideas they find to be of interest or worthy of debate, regardless of which observations are the most recent. Subjects are discussed in scattered fragments rather than coherent wholes. The medium allows every user to turn to his or her point of interest and engage in a unique way with the debate based on his or her own interests and mental associations. A useful metaphor for this mode of social and intellectual engagement is the kaleidoscope. The Internet kaleidoscope reflects each fragmentary contribution as its users turn from debate to debate. The medium does not bring its fragments into a single coherent whole, whether we take this "whole" to mean Habermasian rational discourse or a realist master narrative. It is possible, however, to discern the organization of an Internet discussion. One way to do so is by constructing a tree diagram, and this method will be employed in this chapter to understand how a single Internet discussion branches out to engage with various topics and perspectives. Those who delight in Internet communication do not attempt to absorb the entire "discussion" but enjoy finding their own order in those comments they happen to read.

To recognize a medium like the Internet as a foundation of the public sphere requires that we break from Habermas, not only as Douglas Kellner argues (2000) by discarding his overly rigid dichotomies (classical versus contemporary public spheres, system versus lifeworld, production versus interaction) but also by recognizing that political opinions can be developed through communicational engagements that are too kaleidoscopic to be considered "rational debate." Thus to appreciate the Internet's role in forming the public sphere we must adopt the Habermasian notion of unconstrained access to political debate, but loosen up the boundary conditions on what constitutes debate, including non-rational (though not necessarily irrational) communications such as images, humorous video clips, and music, as well as tangential or "kaleidoscopic" forms of logical association. Habermas leaves room for this broadening by noting that communication depends simultaneously on objective, subjective and intersubjective validity claims, and therefore even subjective communication (like jokes or aesthetic preferences) can have a rational dimension if they relate in a sincere way to subjective conditions or in an ethically defensible way to intersubjective conditions (Habermas 1984, 8–23; 1987, 120–27).

Thus, the virtual space of the Internet provides a context for political discourses related to communities – physical and virtual, as well as bridgespaces between communities. It also records and preserves these comments (for a frustratingly short period, but nonetheless for some period) which allows for the observation and analysis of interactive, casual, grassroots political discourses as I have done. Online communication contexts permitting political discussion and debate take three basic forms: *discussion forums* in which members (who have provided a user name and password) are free to start a discussion or add their comments to a discussion that has been started by another member; *blogs,* in which a single individual initiates many discussions by posting short essays, observations, or questions at regular intervals then overseeing the posting of online comments; and *chat*, which is text-based interactive, real-time discussion. Only the first two of these will be considered here, due to time constraints during the gathering of the data.

The Election on Discussion Forums

Two discussion forums are considered here, one hosted by France's leading newspaper, *Le Monde,* the second hosted by Wanadoo, the French equivalent of AOL (American Online). Very little control (as through censorship or exclusion of certain participants) occurs over these spaces, so it is not expected that *Le Monde*'s forum will necessarily bear any connection to *Le Monde*'s editorial views, although some self-selection of participants by political perspective may occur. Thus we might expect *Le Monde's* contributors to generally take a more middle-of-the-road stance on political issues. Wanadoo is an Internet portal providing various services to its members and some services to nonmembers, as well. Members can contribute to Wanadoo forums while non-members can read but not contribute to the various discussion forums hosted by the portal.

While the two forums varied somewhat in format, they shared some common traits: numerous discussion threads created by an initial posting, several initial responses, responses to those responses, and so on. Perhaps a more accurate word than "thread" would be "tree," since the structure of the communications in Internet discussion forum is tree-like, with a trunk provided by an original posting and both long and short branches consisting of replies to the original posting and subsequent replies to those replies. Figure 7.1 shows one such tree-like conversation structure in diagrammatic form, indicating only the screen names of participants and the general topic of their contributions. From this diagram it is evident that some responses to the original posting (like that of Clairville near the top and just right of center) create major branches while others (like that of "calder" near the top and just left of center) become dead ends. Users navigate through the discussion by choosing particular postings or by viewing postings in chronological or logical sequence.

Quantitative analysis

Four discussion threads were chosen for the quantitative portion of the study. The four threads were all relatively long. The longest contained 152 postings – an

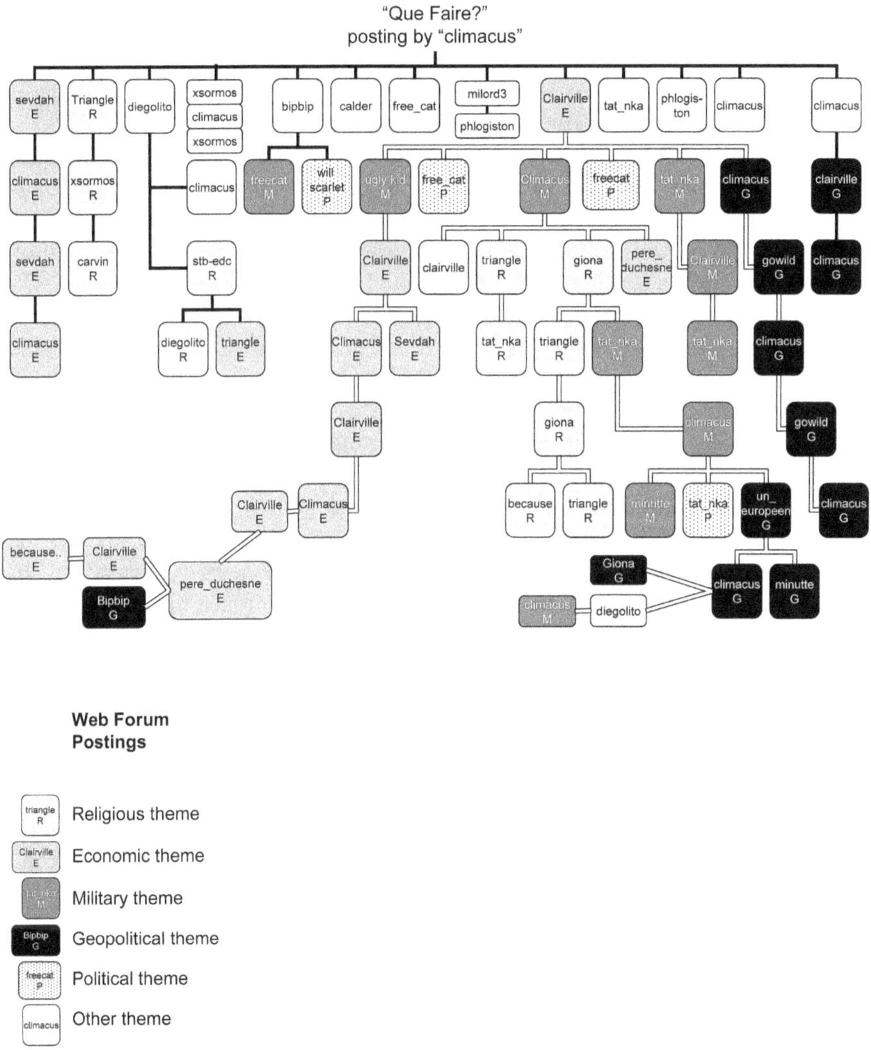

Figure 7.1 Tree diagram of a discussion forum based on a posting of November 3, by a participant using the screen name of Climacus.

extraordinary number – while the shortest contained only 29. Approximately three quarters of the threads in a typical forum contain less than 20 postings, so even the shortest thread considered here was relatively long. The total count of postings considered here is 367. In general, the longer the discussion thread the more it tends to digress and therefore the less influential the original argument or question in shaping the mix of opinions in the thread. By studying long discussion threads it is therefore possible to gain a sense of the prevalent attitudes and interests of forum participants independent of the original poster's perceptions or representations, particularly if views are measured in terms of the relative number of participants expressing a view rather than the number of postings that take a particular view. Two of the discussions were on *Le Monde*'s discussion forum (www.lemonde.fr) and two were on the Wanadoo discussion forum (www.wanadoo.fr). In both cases the discussion was dedicated to the US presidential election.

The postings were analyzed for attitudes relative to Americans, George W. Bush, France, and Europe. The level of analysis was basic, simply discerning which postings took positive or negative stances on each of these four issues. Positive or negative portrayals of Americans might deal with Americans in general, American culture, particular Americans (excluding the presidential candidates) the US economy, US politics, US foreign policy, etc. The "pro" position included any favorable mention of any of these subjects while the "anti-" position included any critical or pessimistic mention of any of these subjects. To qualify as relating to George W. Bush, a posting was required to contain a positive or negative assessment of *either* presidential candidate, which was then interpreted such that favorable comments about John Kerry and unfavorable comments about George Bush were coded as "anti-Bush," while favorable comments about George Bush and unfavorable comments about John Kerry were coded as "pro-Bush." To qualify in the pro-France or anti-France category a posting was required to contain a positive or negative assessment of French culture, French government, the French economy, French historical figures, or any other aspect of French society and culture. The majority of the "anti-France" postings were critical of the Chirac government or critiqued government policies in France. Finally, to qualify as "pro-Europe" or "anti-Europe" a posting was required to include either favorable or unfavorable mention of Europe, the EU, European government, European economy, European policies, and so on.

The Threads On November 3 and November 4 very similar postings appeared on Wanadoo and lemonde.fr. A participant with the screen name of NON A BUSH (NO TO BUSH) started the *Réélection de Bush* (Reelection of Bush) thread on Wanadoo and a participant with the screen name of Climacus started the *Que faire?* (What to do?) thread on lemonde.fr. In both cases, the initial posting argued that the French should respond to the re-election of George Bush by isolating the US through a boycott of US currency and goods. It is certainly possible that both postings were by the same person, as the screen names NON A BUSH and Climacus could easily have been created by the same user. In both cases the call to boycott the US prompted an unusually protracted exchange of views on wide-ranging topics. On Wanadoo one hundred and forty-eight comments were posted on the thread while on lemonde.fr seventy-one comments were posted. A comparison of the two discussions with regard to positions expressed on George Bush and the United States (or Americans) reveals a somewhat different mix of perspectives (Figures 7.2 and 7.3).

The Wanadoo discussion was dominated by critiques of George Bush. Forty-five percent of the participants expressed negative opinions of the American president ranging from dislike to abhorrence to fear. Only twenty percent of the participants expressed positive views of Bush. Counting the number of postings rather than the number of participants slightly reduces the gap between these two positions, but the anti-Bush perspective remains almost twice as common. In lemonde.fr the consensus was even more striking. While less than five percent of the participants approved of Bush or supported his policies, more than twenty-seven percent of the participants disapproved. The ratio of non-supporters to supporters was six to one. Comparing postings rather than participants, the ratio is still close to this level of opposition to Bush (5.5 to 1). The defenders of Bush were seriously outnumbered by the critics in both forums although *Le Monde*'s online community was more strongly opposed to Bush.

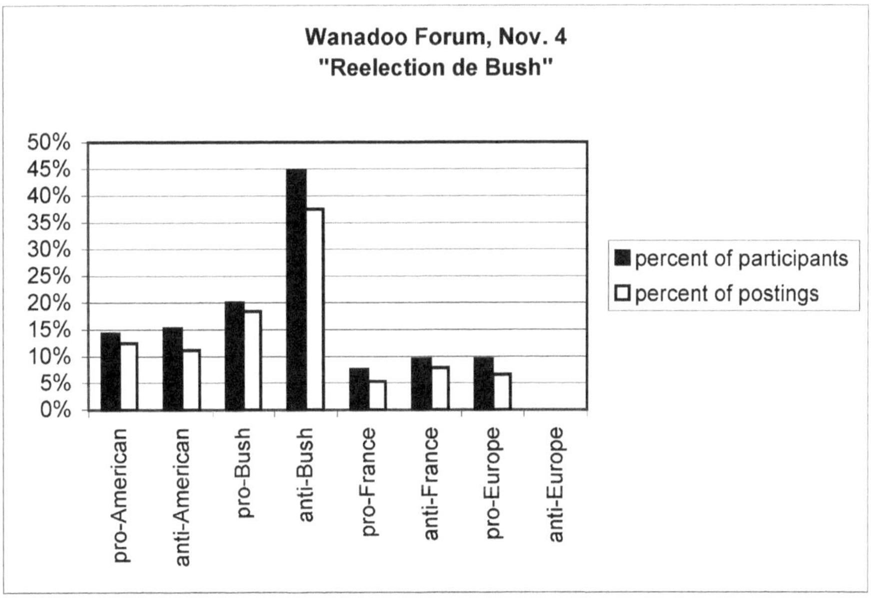

Figure 7.2 Analysis of opinions expressed in a November 4 forum on www.wanadoo.fr: "Re-election de Bush." Percentages are shown for postings and participants adopting "pro-" or "anti-" perspectives with regard to the United States, George W. Bush, France, and Europe. "Pro-" perspectives express support, interest, or approval, while "anti-" perspectives express fear, dislike, criticism, or other negative positions.

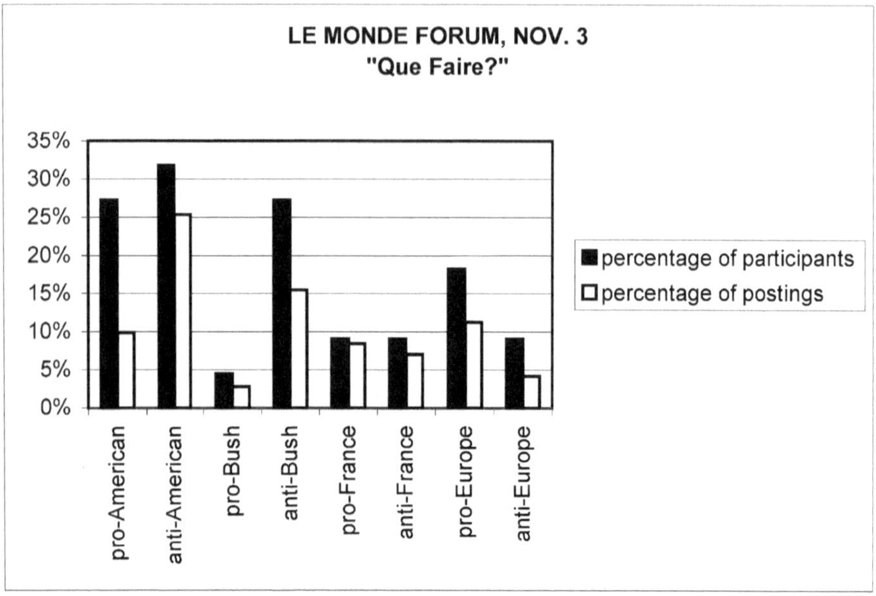

Figure 7.3 Analysis of opinions expressed in a November 3 forum on www.lemonde.fr. Percentages are shown for postings and participants adopting "pro-" or "anti-" perspectives with regard to the United States, George W. Bush, France, and Europe. "Pro-" perspectives express support, interest, or approval, while "anti-" perspectives express fear, dislike, criticism, or other negative positions.

The difference between these two discussions is also manifested in regard to expressed views of the US. The *Le Monde* discussion gave greater attention to the US apart from George Bush, with sixty percent of the participants expressing some sort of opinion about the US. On wanadoo.fr less than twenty-nine percent of the participants expressed positions relative to the US. Strikingly, on both forums, the pro-American and anti-American stances were almost evenly balanced in terms of the number of participants.

Because of the principle of the noisy outlier, it is necessary to measure opinions by number of postings *and* by number of participants. It could be that two fifths or even half of the postings in a forum support a certain idea but only one person accounts for virtually all of the postings taking a particular stance. For this reason we can see why the "pro-Bush" position is more prevalent when measured by the number of postings than by the number of participants. This is because supporters of Bush, sensing themselves in the minority, became hyperactive in asserting their views. However, if we judge by "percentage of participants" then both discussions demonstrated strong disapproval towards the American president, with the Wanadoo participants opposed to Bush by a ratio of more than 2:1 and the *Le Monde* participants opposed to Bush by more than 5:1. We naturally see much less interest in other issues except in the case of *Le Monde*'s forum and the issue of the US. Here, as noted before, the opinions are balanced. Opinions on France were similarly ambivalent but opinions on Europe, albeit numerically limited due to the focus of the discussion on America and the American election, were quite strong. Measured in terms of postings or participants Europe benefited from strong support. The symbolic link between a strong and threatening America, *l'hyperpuissance*, and a need to unify Europe and create *Europe-puissance*, is therefore clearly indicated. One intertextual source of this connection is scholarly writing (chapter 4), although it resonates to some degree in other media such as television and newspapers. Interestingly, the authoritative communication of scholarship is more strongly reflected in the kaleidoscope of the Internet forum than on television or in the newspapers.

The link between anti-Bush and pro-European sentiments is revealed more clearly in figures 7.4 and 7.5. These charts measure the frequency of pro- and anti-sentiments relative to each theme in the threads. Interestingly, when opinions are measured in this way, and four separate discussions are included in the calculations, counting postings and counting participants produces virtually the same opinion profiles. The forums included in the above comparison, in addition to *Que faire?* and *Réélection de Bush*, were one originating with a posting called *Quand la Chine supplantera les U.S.A.* (When China replaces the USA) on Wanadoo, and one from lemonde.fr called *Moi, je vais voter pour Bush* (*I'm* voting for Bush). The

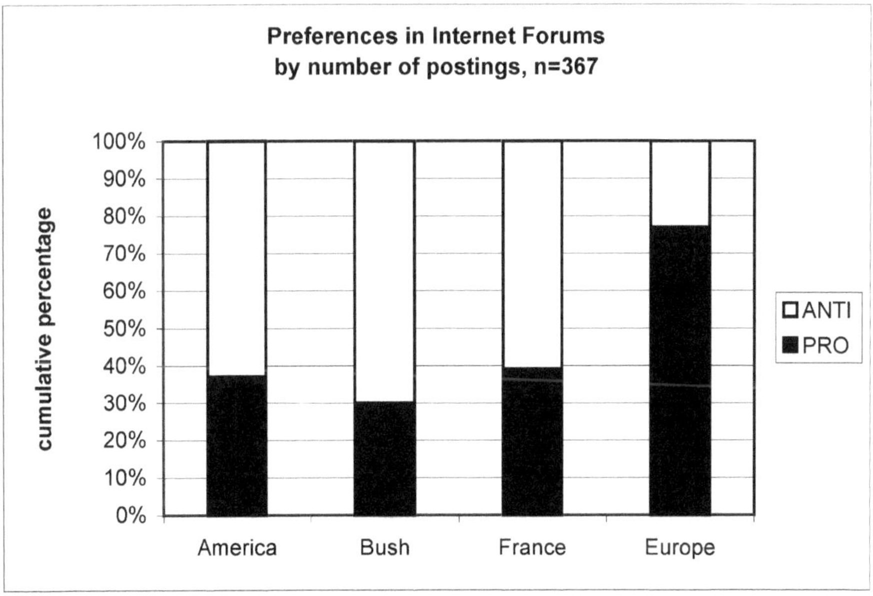

Figure 7.4 Cumulative percentages of all positions on four issues, in four Internet forums–two on www.lemonde.fr and two on www.wanadoo.fr. Calculated on number of postings in which the positions are expressed.

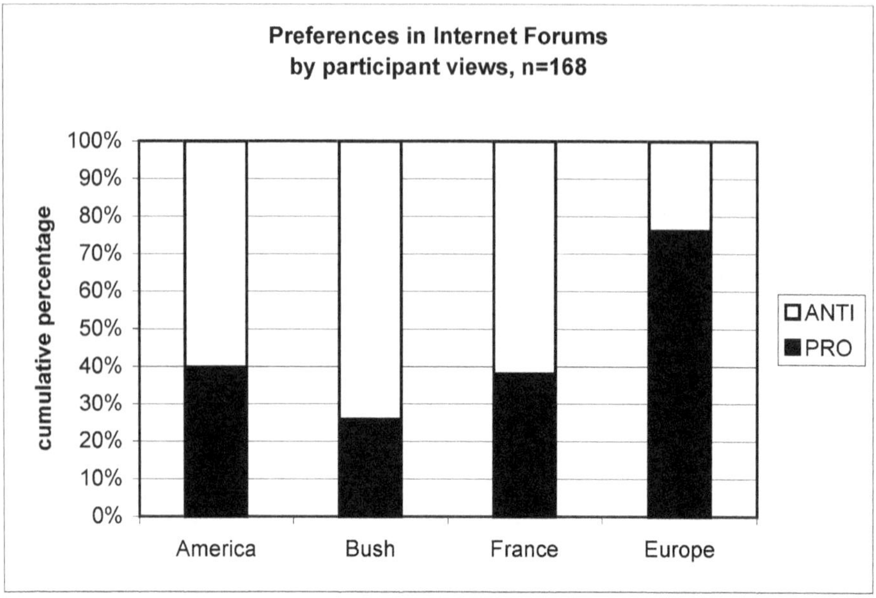

Figure 7.5 Cumulative percentages of all positions on four issues, in four Internet forums–two on www.lemonde.fr and two on www.wanadoo.fr. Calculated on number of participants who ever express a given viewpoint.

subjects of the latter two discussions, respectively, were the weakness of the United States (*Tottering Giant* motif) and the differences or lack of difference between the candidates (*Big Difference* and *No Difference* motifs). Predictably, these threads generated considerable discussion about the US, but the opportunity to indulge in anti-Americanism seems to have been no more attractive than the opportunity to criticize the French government. Bush receives only 25 percent to 30 percent support; both France *and* America receive close to 40 percent support; and Europe receives just over 75 percent support. The total count of postings and participants for all four forums combined can be seen in figure 7.6. Here the critique of George Bush is strongly indicated. The mix of opinions in the other forums alone can be seen in figures 7.7 and 7.8.

The quantitative analysis has indicated considerable engagement by certain individuals with views that diverge from the norm of a given discussion and I have indicated their role by the term "noisy outlier." One other predictor of participation level is whether someone is the originator of a thread, a status which tends to encourage frequent and sustained participation in the thread. This is easiest to understand by referring to the prevalence of Climacus in figure 7.1. After originating the thread, Climacus proceeded to post 17 out of 70 postings, or close to a quarter of all of the postings in this discussion. Climacus could be described as an

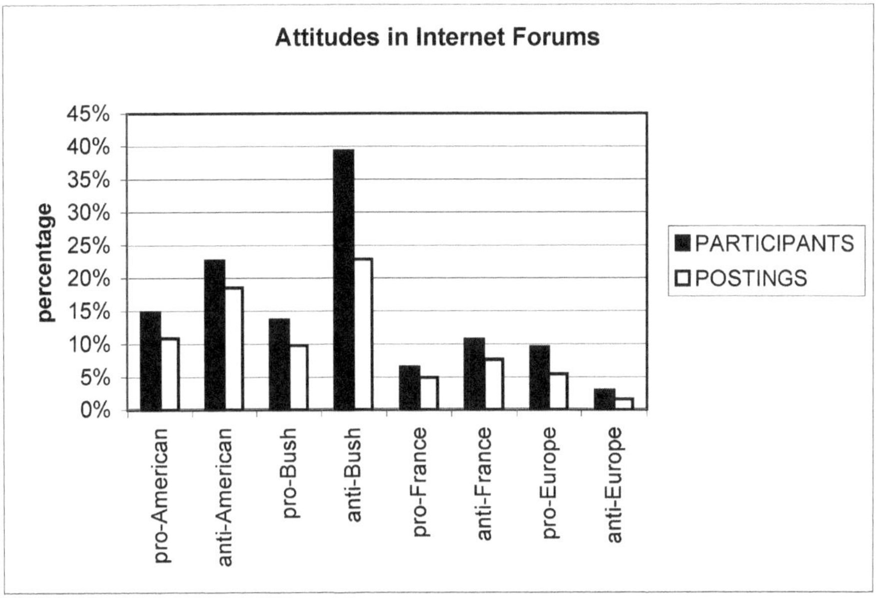

Figure 7.6 Analysis of opinions expressed in four Internet forums–two on www.lemonde.fr and two on www.wanadoo.fr. Percentages are shown for postings and participants adopting "pro-" or "anti-" perspectives with regard to the United States, George W. Bush, France, and Europe. "Pro-" perspectives express support, interest, or approval, while "anti-" perspectives express fear, dislike, criticism, or other negative positions.

"active initiator." The combination of the "noisy outlier" and the "active initiator" contribute to an interesting relationship between one's ranking among participants (based on the frequency with which one contributes) and the number of times one participates. Those who participate only once or twice are quite numerous and those who participate a great number of times are uncommon. In between we see the characteristic curve of a power law, much like the rank-size distribution of cities (the Zipf law) or the connectivity distribution of the Internet itself (Barabási 2003; Buchanan 2002). Figures 7.9 and 7.10 demonstrate this interesting property in two of the discussions analyzed here. In summary, we can see that criticism of George W. Bush and support of Europe are strongly evident while this tension is not reflected in a nationalistic or anti-American consensus.

Qualitative analysis

Keeping in mind the quantitative findings of dislike towards George Bush, support for Europe, remarkably similar levels of criticism relative to the US and France, and vastly disproportionate levels of participation among participants in a given

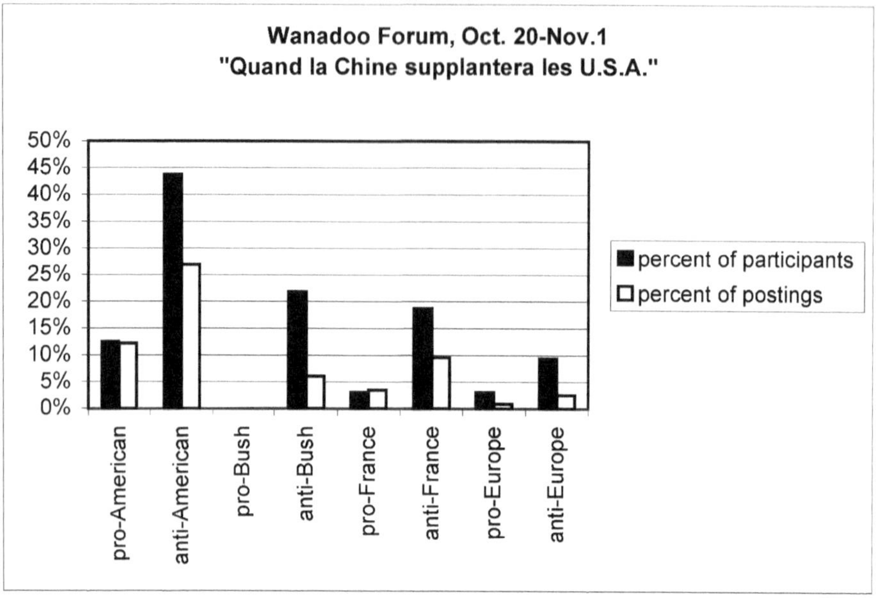

Figure 7.7 Analysis of opinions expressed in an October 20 to November 1 forum on www.wanadoo.fr. Percentages are shown for postings and participants adopting "pro-" or "anti-" perspectives with regard to the United States, George W. Bush, France, and Europe. "Pro-" perspectives express support, interest, or approval, while "anti-" perspectives express fear, dislike, criticism, or other negative positions.

discussion, we can move on to look at the unique aspects of the topics raised and rhetoric employed in these discussion forums. Only by pursuing the discussion and debate in particular threads/trees can we begin to flesh out the impressions from the quantitative analysis with a sense of how the Internet served people as a virtual context in which to redefine citizenship, nationality, and political culture. Unfortunately an orderly extraction of motifs is not as easy with the Internet as it is with the newspaper articles or television programs, because of the attention deficit and the astounding range of opinions and topics woven together by an Internet forum. Beyond pro-Bush and anti-Bush arguments, a listing of the "most important" motifs is not possible. Consequently we will discover motifs only as we find them embedded in particular postings.

Now it is time to engage more directly with these discussions by following one through from beginning to end, then by looking at some key points in three others. My objective in doing this is to reveal themes (just as I have throughout the book) but now the focus is on *connecting themes* rather than *structuring themes.* How does a comment about George Bush lead to a comment about the European Union?

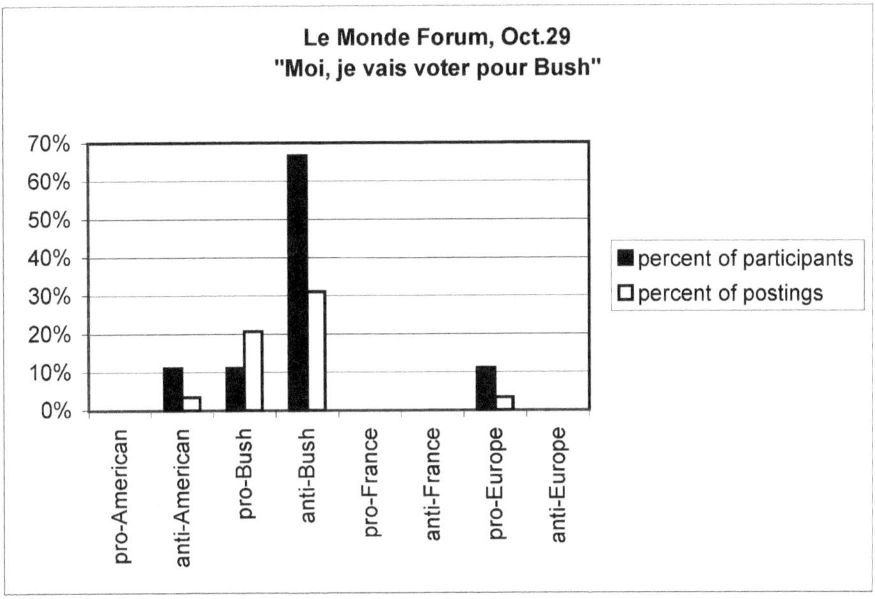

Figure 7.8 Analysis of opinions expressed in the Internet forum "Moi, je vais voter pour Bush" (I'm going to vote for Bush) 29 October 2004, www.lemonde.fr. Percentages are shown for postings and participants adopting "pro-" or "anti-" perspectives with regard to the United States, George W. Bush, France, and Europe. Pro- perspectives express support, interest, or approval, while anti- perspectives express fear, dislike, criticism, or other negative positions.

What are the particular twists and turns people take in the discursive space of the online debate?

As noted above, items posted on the Internet have a linguistic style that resembles unpublished writings like casual notes and personal letters. I have tried to capture this informality in my translations of the French Internet postings by omitting capital letters where the original did so, and by copying the punctuation (or lack thereof) as closely as possible. Where mistakes were made in grammar, I have tried to convey the flavor of those mistakes in English. Where meaning was difficult to ascertain from the words provided, whether in French or English, I have added information in brackets to clarify the posting. I have also tried to reproduce the linguistic register (the degree of formality or informality) of the original. So if French slang was used, I employ English slang although the literal meaning may be rather different. If formal French was used I employ formal English. In short, every attempt has been made to match the tone and register of the original. Some postings include words or passages in English, and these have been indicated explicitly in my translation. The one aspect of the original that has not been preserved is the arrangement on the page,

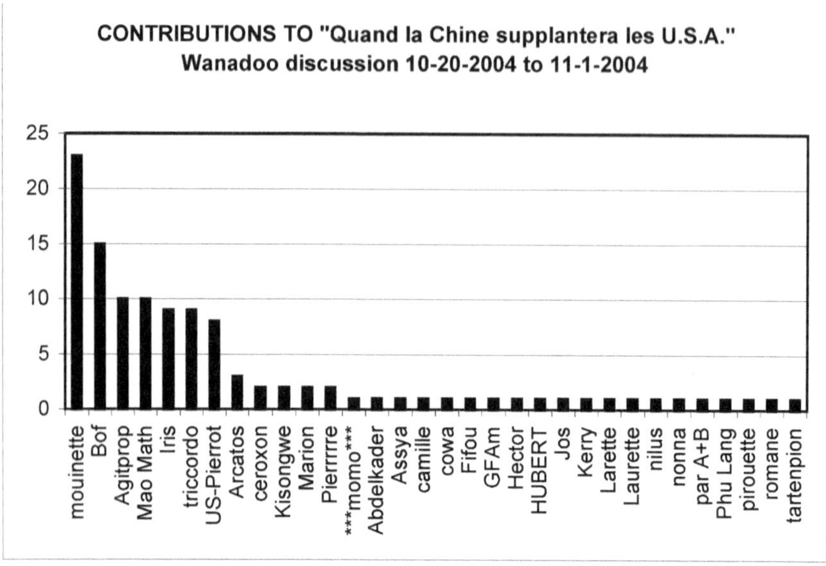

Figure 7.9 **Frequency of participation by 32 participants in the Internet forum "Quand la Chine supplantera les U.S.A." (when China replaces the US) 20 October to 1 November 2004, www.wanadoo.fr.**

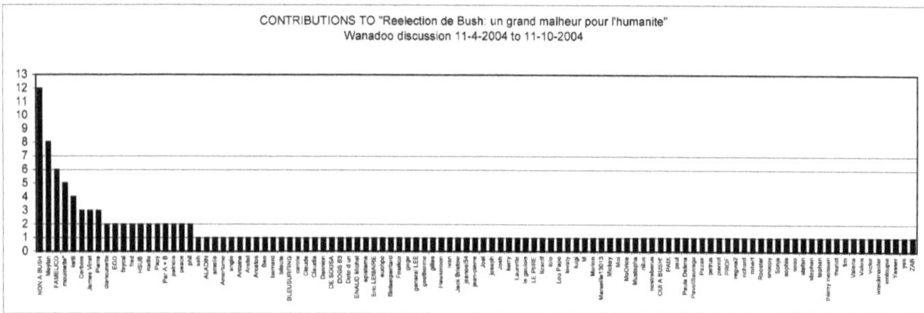

Figure 7.10 **Frequency of participation by 105 participants in the Internet forum "Reélection de Bush: un grand malheur pour l'humanité" (Reelection of Bush: a great misfortune for humanity) 4 November to 10 November 2004, www.wanadoo.fr.**

because some participants insert extra returns between sentences that do not affect the meaning of their contributions.

Moi, je vais voter pour Bush An October 29 discussion on *Le Monde*'s online forum demonstrated the principle of the noisy outlier as well as a remarkably seamless joining of participants in France and North America. Although it is difficult to ascertain the exact locations of participants, evidence indicates involvement by

American, Canadian and French citizens. The discussion thread was started by gowild, who eventually represented herself as a woman living in the US in a swing state (although misrepresentation of personal identity is always easy on the Internet). The title of the original posting was, "Moi, je vais voter pour Bush," meaning "*I'm* voting for Bush."

The discussion that ensued from her comment included many bilingual postings in which English was used more comfortably than French. Certain participants, most notably gowild, had difficulty expressing themselves in written French, made numerous mistakes in grammar and spelling, lapsed into English to express figures of speech, and included quoted segments in English. These linguistic clues, along with self-presentation or presentation of others, all indicated that several participants resided in the United States. Gowild's errors are indicative of someone who has studied French in school but has not spoken French very often, for example writing "ma opinion" instead of "mon opinion," which is unnatural in spoken French the way "a apple" is unnatural in spoken English and indicates someone who has not lived in a Francophone environment, yet the large vocabulary of this participant indicates someone with an extensive education in French or else a French spouse or partner. One participant, kadiatou, used the term "états uniens" for Americans, which is indicative of Québécois French. Other participants wrote in fluent and casual French without indications of Québécois dialect, indicating that they were French citizens situated in France. The signs therefore pointed to a political debate linking citizens of at least three nationalities in a single "debate."

Gowild started the thread by writing: "Good day or rather good evening!" The time marked by lemonde.fr was 1:38 am but this would be 7:38 pm in the eastern US, and 8:38 pm in the central time zone. Later in the discussion she (for ease of discussion I will accept the participant's self-presentation of gender) revealed her location as one of the swing states, so she had reason to say "good evening" rather than "good morning." She continued in clumsy French: "Don't shoot please. I know, I know, he is an idiotic (*abruti*) cowboy from Texas … but I prefer him better than Kerry. Kerry is very frightening. So it's the same old thing, it's the lesser evil. But I must say that I detest his point of view on France. As for the rest he is right …"

Another participant using the screen name "because..." quoted gowild's observation, "Kerry is very frightening," and simply asked "In what?" Gowild replied to this, "many reasons… a vile record on taxes, lack of credibility, integrity of character… there is no clear vision on foreign policy… etc. etc. Slavic also replied to gowild, asking: "and what do you find that is so positive in Bush's camp? how do you think he has succeeded? Do you think Bush's bottom line (*bilan*) has been so productive for America? The Reagan and Bush One periods were fruitful for their country, even if exacerbating [America's] Reagan style (the B-rate movie cowboy), [while] bush junior is in the process of plunging his country below all other cultures (*plonger son pays sous ttes les coutures*), at least that's what we learn and the impression it leaves." To this, gowild replied defensively: "Bush is a true guy not false like Kerry. Maybe a bit stubborn, true, but a man who does [i.e. acts] rather than speak[ing]. Bush took [over] his country in the midst of a post-Clinton dive (a time when the economy was in the midst of falling, too – technology bubble burst). Then September 11 arrived… These are very serious affairs for one to handle. …and this is a very serious discussion for me to continue in French but I still do not abandon it."

Nevado copied gowild's passage starting: "a vile record on taxes, …" and attached a comment addressed to other participants in the discussion forum: "I hope you are all impressed by the weighty arguments of the Bush supporters (very well represented by gowild). That's really feeble (*Ça vole bas*)!" This posting broke the pattern of one-to-one communication established up to that point and addressed others who might be following the discussion without posting comments (the "lurkers," in Internet parlance). The posting was written in fluent French suggesting that Nevado resided in France or at least spoke French as his/her mother tongue. Nevado's sarcasm and the shift to address lurkers had two effects on the terms of debate. First, it initiated a move away from the presidential candidates and toward the participants themselves as subjects of discussion. Second, it rejected the one-to-one discussion up to this point on the basis that it was lacking in substance and appealed to a broad audience (although without trying to present a counter-argument of substance). This posting therefore marked a break in two forms of order, one related to communication context (a shift from one-to-one to one-to-many) and the other related to communication content (a shift from discussing the candidates to discussing the participants).

Over the next ten minutes a user going by the name of "because…" posted two comments, both incorporating long passages in English that were copied and pasted into the forum from other sources. One was a joke about Bush's intelligence, the other, an analysis of factual distortions in Bush's autobiography. These borrowed English quotes demonstrated the intertextual and polyvocalic nature of the Internet. The borrowings added complexity to the Internet as a medium, indicating ties to other media such as books and television and ties between English and French language communities. This contributor also took an ironic perspective on American politics, an attitude considerably more common in France than in the US. He/she noted that Bush is a limitless source of comic material, that eight years of Bush would prompt people to vote for Hillary Clinton in 2008, and that Tony Blair's troubles demonstrated the difficulties faced by European politicians supportive of Bush. The original (non-copied) portion of the posting ended with a comment that stood alone without supporting arguments or evidence but compellingly reflected the findings from previous chapters: "it will be way easier to be pro-european with bush."

At this point Clairville contributed two postings. These were the most bizarre contributions to the discussion and they accentuated the previously mentioned breaks in the pattern of communication context and content (from interpersonal to group appeals) while adopting an abusive attitude and maintaining the focus on the discussion participants rather than the presidential candidates. The first read: "Vote for Nader; you will be doing a good thing and God will repay you. It's important God and he isn't the patron of a whorehouse, *house of ill repute*, like Satan-Bush believes. *Repent sinner!*" (grammatical errors in the original version; italicized phrases in

English in the original) to which gowild replied mildly "Thanks for your advice but I doubt it will improve my image before God." Clairville's second contribution read: "Yes, but Bush is a Zionist, and his great grandmother born in Brooklyn is a Jew. Must not exaggerate! *Sir, are you a kike*?" (Italicized phrase in English in the original). Again gowild de-escalated the antagonism, commenting: "maybe, but he's not alone. All of America is Zionist! and also I am not a 'sir' and I cannot find 'kike' in my vocabulary… very sorry (I don't need a translation if this is an insult..)" In the absence of control or moderation, the discussion had drifted toward an exchange of flaming and defensive comments rather than a real debate with contestable claims. Interestingly, it is at this point that the gender of one of the participants was revealed. Along with conflict came an impulse to *appear* in some manner "I am not a 'sir'" and not just to assert a political position. While trying to restore civility and rationality gowild used personal information and incomprehension together, as discursive tools.

Subsequently gowild responded to Nevado's postings in French, although clearly she and Nevado were both Anglophones. "and you represent all French people of course? However I am a dummy in French [to] give you many acceptable reason[s] but my point of view is closer to Republicans than Democrats. Why do you think Kerry would do a better job than Bush?" Nevado did not reply until thirteen hours later, but Clairville chimed in much sooner with the following, posted in English: "Mr. Nevado is a jerk from California. He's a commie and should be in jail. His former boss was Gus Hall." To this gowild replied rather weakly, "I thought you were asleep already."

Clairville employs a communication pattern common among certain men on the Internet but rare among women – engaging in confrontational rhetoric for the sheer delight of irritating and annoying other participants (Herring 1999; 1996). Despite the use of politically charged words like "kike" and "commie" Clairville's postings have as much to do with the personal gratification of annoying others as with politics. In the context of political communication this kind of behavior works against an ideal speech situation, instigating conflicts rather than attempts to reach common understanding. That this anti-democratic communication was interwoven with English might have led rather quickly to expressions of anti-Americanism among the other participants but it did not, probably because most of the participants at this time of the day (after midnight in France) were Americans or French expatriates living in the US.

Gowild's last comment drew attention to the time of day of the discussion. It was 4:37 am in France according to the record on the posting, but on the east coast of the US that would have been 10:37 pm. The time of day appeared to be shaping the content and language of the forum. The forum was not simply declining into rudeness and flaming, but it was doing so at a time when the French, who approached the election somewhat more philosophically than Americans, were generally asleep. But if the participants appeared comfortable with English, confirming the impression that most of them were actually in North America, the discussion nonetheless remained in the French language (although the grammar was of varying caliber) and the predominance of North Americans among the participants was not discussed up to this point. The next posting was therefore a bit of a departure.

Clairville replied to gowild: "Being in the city that never sleeps, I do same. Sometimes I doze off," to which she asked: "let me guess – New York, LA?" The comment was meant to "out" Clairville, geographically speaking, but it was met with digital silence.

At this point Tat_nka joined with the comment: "Hi Nevado, I really hope for your country that the majority does not fall for this [*ne s'en tient pas à ça*]! If not, you're off to a bad start!" The "your country" comment alludes to the assumption that Nevado is in the US, although this assumption is not confirmed by Nevado. What is less clear is the meaning of "this" in the posting. Does "this" refer to Nevado's anti-Bush perspective or the deceptions by Bush that are documented in the material Nevado has pasted into his posting? Such ambiguity plagues Internet discussions. Insofar as ambiguity leads to unexpected turns in a discussion however, it might be said to enhance rather than plague Internet discussions.

Gowild now replied to a much earlier posting in which "because…" copied evidence that Bush twisted a family story to serve political purposes: "it's just a thing for his [Bush's] trademark image and facts are true facts only in math … in politics they can always be twisted by one side or the other …"

After this apology for Bush, the postings stopped for a period of ten and a half hours, a period corresponding to the interval from 6:00 am to 4:30 pm in France and from midnight to 10:30 am in the US. In fact the gap is best understood spatiotemporally as the interval between midnight on the US's east coast and 6:00 pm in France, the space between two evenings an ocean apart yet joined by a single conversation. Three out of five participants joining after this interval are new and are clearly French, indicating that the discussion had now shifted continents from North America to France (at least with regard to most participants). What was before primarily a North American debate among people pretending to be French (but occasionally "outing" each other) is now primarily a French debate with only a few North American participants. Most of the remaining comments are posted between 5:00 pm and 10:30 pm in France (11:00 am and 4:30 pm on the east coast of the US).

MacGee broke the 10 ½ hour silence, writing to gowild: "Sure you're going to vote for Bush because you judge that Kerry is 'frightening' but you don't explain in a very precise manner what it is about the Democratic candidate that makes you fearful and what attracts you to the Republican ticket. In addition, if you would permit me, do you live in a state where the game is already played out or are you in one of these infamous 'swing-states'? Is your primary concern about the security of your country as well as the anti-terrorist fight or is it at the level of economic risks encompassing very particularly the question of employment?" This carefully written question tried to steer the discussion away from personal slights and insults and toward rational discussion. The effort to reframe the content and context of discussion did not go unnoticed. Bvsc remarked: "I am amazed he can find people who are sufficiently civil and polite [*gentil*] to respond…" What is ironic about this comment is that it appears supportive, but as it is constructed so as to speak *about* MacGee rather than *to* him/her it does not actually help return the discussion to a rational plane. Gowild, who was amazingly still online, noted this comment and wrote haughtily to bvsc: "Me too, I am surprised you do not accept that my opinion could be different from yours but we could still have a dialogue." Nevado cut in, responding to gowild's comment to bvsc and wrote snidely: "You are confusing opinion, which flows [*découle*] from reasoning, and credulity, into which people fall who are incapable of reasoning." Gowild's response to MacGee's carefully crafted contribution was peculiar: it took the approach of presenting a carefully considered list of debating points while at the same time denying an interest in politics. "Kerry is very Left, very liberal. His record is stuffed [*truffé*] with flip flops (trade with China, support for troops in Iraq and its corrupt leader [*succube*] under pressure from Howard Dean), bad idea[s] concerning affirmative action. Bush has a good record on judicial nominations, he has [demonstrated] greater government responsibilities, a clear position on pro-growth policies and Jobs for the 21st Century Initiative. I have no interest in his point of view on religion, his support for the NRA and [opposition to] stem cell research. But like I said before, he is the lesser evil. And yes I live in one of the swing states. But politics are not a question of life and death for me. Bush or Kerry? Life will not come to a halt in any case." This odd mixture of political passion and feigned apathy presented like a lesson to other participants caused exasperation.

One was bvsc, who exclaimed: "How can one vote for an inveterate liar, an ex-alcoholic, a coward, a medieval religionist, a stupid and uncultured Daddy's boy who keeps repeating the same untruths all the time, who every day causes men of his own and other countries to die, [and] civilians and children, solely for the interest of the big companies that financed their campaigns?" The above was written in French – sloppy French but clearly native French. The concluding words were in English: "You make me sick, I give up." The linguistic switch appeared to indicate condescension toward gowild, as if her political naiveté were based on misunderstanding of the language.

MacGee remarked deviously: "I agree very much with your conclusion, our lives will not stop whether it's Bush or Kerry who is elected. That much said, some of the points in your comment bug me [*me turlupinent*]. You claim to have appreciated the judicial nominations carried out by Bush. I suppose you are talking about federal judges. But all the same they have a very partisan flavor (the majority of judges named have clearly conservative positions) while this same Bush promised to reunify the nation and not divide it. All the same, if he is elected he will name two or even three Supreme Court justices who will certainly have to decide on important subjects such as the right to abortion and cases on individual liberties (already blunted by the Patriot Act). Are you sure that totally satisfies you? As for his employment plan, I've noticed that during his first mandate poverty grew in the US and more jobs were destroyed than created. In addition, his tax cuts have above all favored the most privileged classes and he has greatly increased the budget deficit. What can his miracle recipe to fix the situation really be if he has a second term? Even so, I don't think W. will go so far as to suppress affirmative action; what distinguishes him on this level from Kerry? As for "flip-flops" of the Democratic candidate, permit me to send you to an editorial from *The Economist*: between incoherent and incompetent [candidate], you [should] choose incoherent."

Yet another participant responded to gowild with exasperation. In regard to the assessment of Kerry's political position, Nevado asked "Very Left of what? Of the extreme Right?" Then in response to the comment that the outcome of the election is not a matter of life and death, he commented grimly, "It has been, for the tens of thousands of Iraqis (they are talking of 100,000 currently) and for 1,100 + Americans. Even you should watch out."

What seemed to annoy other participants most was the flippancy with which gowild treated the electoral choice Americans would make, an attitude that provoked anger because of the democratic deficit experienced by Europeans (see chapter 1); American presidents make decisions that shape the fate of non-Americans so for Americans to remain blasé about their decision of a president and their own role in the world is a form of dismissal of the needs and even the humanity of others.

Meanwhile, Nevado took up a question posed thirteen hours earlier by gowild. The question read: "Why do you think Kerry would do a better job than Bush?" In reply Nevado wrote, "It's because he [Kerry] has a minimum of grey matter between his ears… and ears that work, as well. That's already not bad. But I understand very well that you feel yourself to be closer to Bush." If gowild noted this subtle insult of her intelligence she did not let on.

Finally a new participant, kadiatou, chimed in with a comment more clearly anti-American than anything else in the discussion: "That's good [that gowild prefers Bush]. Useless to shoot at you because you are no longer dangerous. One more well-programmed American [*un états uniens* (sic) *de plus 'bien' formatés*]. Go raise your little ones like that: permanent fear of aggression, fear of the axis of evil…? Pity!!! You are anaesthetized." This was a Canadian participant, judging by the term "états-unien" rather than "Américan," the latter term being preferred by the French. The Canadian reframed the discussion in nationalistic terms rather than in terms of the two candidates or the participants in the Internet forum. Gowild had succeeded in confirming at least one Canadian's stereotype of the indoctrinated American.

The discussion was close to ideal in the Habermasian sense. Participants were free from internal and external constraints. Yet it mainly served to demonstrate disagreement rather than resolving it. This discussion is analyzed in Figure 7.8 with regard to the US, Bush, France and Europe. Over one third of the postings (11/29) were by one person and she (he?) took a political position opposed to that of virtually all of the other contributors. In this case the "noisy outlier" was also an "active initiator," a situation that generated much controversy, though not generally of a constructive nature. Anti-Bush postings outnumbered pro-Bush postings by only 33 percent since there were eight anti-Bush postings as opposed to six pro-Bush postings. But the noisy outlier generated *all* of the pro-Bush comments while five participants generated the anti-Bush comments, so the ratio of anti-Bush to pro-Bush participants was 5:1. The debate constantly escaped the person who initiated it. Rather than exploring the details and the logic of the anti-Bush perspective, this discussion just increased the emotional freight of that position. Three postings were incoherent, displaying the effects of some form of intoxication or mental illness perhaps, or at least an aggressively disruptive attitude. Only one posting was anti-American, and it came from the only identifiable Canadian. Over a quarter (8/29) of the postings were comments about other participants rather than responses to the topic of debate. These postings ranged from "I thought you were asleep already," to questions and speculation about another participant's place of residence, to insults and contemptuous asides. Geography matters in such a "placeless" debate, but not in the way it ordinarily matters. It is transformed into something hidden, subterranean and mysterious in communication content rather than a structuring force of communication context that permits or prohibits communication itself.

At a more general level we can see that the *Le Monde* forum served more than one purpose. For native Francophones in North America it provided an easily accessible political forum in the mother tongue; for English speakers in North America it provided the means to taunt and bait the French (who in the evening hours in North America are, however, scarce to nonexistent on the forum). It is probably somewhat disappointing for American participants on lemonde.fr to discover that they are in fact arguing with and baiting other Americans or at best French expatriates living in the US. The feeling is mutual, so all of the participants are coy about their geographical location. This locational uncertainty preserves involvement in the forum and mutual interest among participants who have deliberately sought out *Le Monde* online rather than North American communication contexts because it seems to offer access to a French public sphere. Socially speaking, the forum may be North American at certain times during a 24 hour cycle and French at other times, but physically (as infrastructure based at *Le Monde*) and symbolically (as a virtual "place" to engage in free-form French language argumentation) it remains constantly in France. This is a hybrid space produced by globalization, a place for political encounters where people are not afraid to speak their mind. But is it a useful supplement to other spaces of political dialogue – mass media and physical place? This remains to be seen. The Habermasian goal of seeking consensus through inclusive, rational, egalitarian debate seems no closer.

Europe Puissance Turning to other strings, true to its title the discussion forum at *Le Monde* captured an international discussion, involving participants in France and elsewhere, and many of the postings strikingly foregrounded a geopolitical vision linking the future of the US to the future of Europe. This vision was a surge of "Europeanist" sentiment in response to the perception of a new wave of US expansion. In a short discussion thread with the title *Et après?* (and afterwards?) the first posting demonstrates the Europeanist logic applied to interpret the Bush victory.

> Four more years of Bush. It's a unique chance for Europe – if Europe takes it. If Kerry had won, he would have pursued exactly the same political program, with the only difference being a "please" and a "thank you." And the European leaders, who have said so much

> in his favor, would show bad form not to support him in his efforts which would have no purpose, in any case, but to affirm the supremacy of the USA over the rest of the world and to defend the interests of his country against Europe.
>
> (D-520)

Here we see the *No Difference* motif linked (negatively) to a *Useful Threat* motif. Kerry, as the less overtly threatening candidate, is therefore less useful. The same day, another participant expressed almost the exact same interpretation of events, starting a different discussion thread.

> You might think the election of Bush is harmful for France and Europe. It's true that the majority of French and Europeans were rather favorable towards Kerry. But it is necessary to offer a number of reservations to the idea that the election of Bush is not good for us. ... [it] will reinforce the positions of France and Germany [against US policy] whereas the election of Kerry would have posed a problem from the moment it was necessary to take a position for or against participation in the war in Iraq. We would certainly have seen a divorce relating to Iraq, the Germans being seduced for sure by the propositions of a Kerry. With Bush we will continue our policies with the hope that even more countries will follow the lead of Spain and join us. From there the alliance, the transatlantic partnership, will shatter into bits and lose nearly all of its meaning. Bush, without intending to do so, cuts the grass from under the feet of the European Atlanticists whereas Kerry would have given them grist for their mill. This is all good for the emergence of a political Europe because it's exactly this partnership that poses problems for us, that prevents Europe freeing itself from the United States, that contributes to making us a political dwarf.
>
> (milord 3)

Responses to these two posts were generally positive, including: "the election ... of Bush is a benediction for Europe," (Tatanka, November 3, 2004). Each of the original posters responded to the others' post: "I'm also of this opinion, but I can't help having a bitter taste in my mouth when thinking of the many sorrows to come for certain ones in this world ... Project for a New American Century... FORWARD MARCH ...EVERYONE!!!!" (D-520); "It's a secular and multicultural Europe that the Bush administration opposes much more than merely France" (milord3). In this last posting, concern with US culture evokes the *Fundamentalism* motif (as in both the newspaper and the television coverage) as well as the *Continental Drift* motif discussed in the previous chapter.

It would be too easy to see the emergence of Europeanist sentiment in Internet forums as simply the coalescence of political identification in a *larger* territory than the existing nation-state, whereas in fact it deviates from territorial definitions of citizenship. The best indication of this is the concluding remark in the thread started by milord3. The contributor's screen name was listed as AnAmericanOpinion and the melodramatic contribution was provided in both French and English (original in English quoted):

> In France. In Europe. Geo-Political Intrigue. No mention of the UN. Demands for more military spending. Strong diplomatic pressure. Ecomonic reprecussions [sic]. Military

> action? Why do the French and Europe do this? For Vanity? No. For Conquest and Treasure. No. They do because their conscience dictates that they must do. So Strong, their passion, their beliefs. It wont be easy but their conscience demands they act! I must shed a tear of joy, for deep down inside they truly are…Americans! God Bless Amer… Europa.
>
> (AnAmericanOpinion)

It is unclear whether this paean is serious or sarcastic. Again we run up against the ambiguity of Internet communication. The Atlantic reverberation is muddled and unclear, perhaps making it all the more intriguing for participants. Assuming for a moment that it is serious, we find here a disillusioned American or Canadian, eager to see in Europe the rejuvenation of democracy but also in fact *participating* in that rejuvenation by "entering" the virtual space of the Internet discussion.

Quand la Chine Supplantera les USA The character of online discussion scarcely differs between lemonde.fr and wanadoo.fr. A particularly long exchange on Wanadoo (115 comments over a two week period) was initiated by an observation about the decline of the United States from a participant using the name ceroxon: "Ten years from now, the way things are going, China will pass up the US in many ways, most notably the economy. The Americans will then have other worries than terrorism. What's more, I never have understood why they want to fight terrorism which only took 3000 lives on September 11 and they do nothing about gangs, which cause more than 12,000 shooting deaths each year in the same country…one time there were only 8 police officers available in the state of Oregon, so why send 120,000 soldiers to Iraq?" The responses to this observation ranged from simple affirmation, to a suggestion that Japan, not China, would overtake the US, to sarcastic ridicule of the idea that the US is vulnerable, to a passionate defense of the US's right to defend itself against "islamo-fascists," to an articulate pronouncement of the link between poverty and terrorism.

The most active participant in this online forum, "Mouinette," accounted for an impressive 20 percent of all contributions to the discussion, more than half again as many comments as from the next most vociferous participant (see Fig. 7.9). Furthermore, she (again apparently a woman) staunchly defended American culture, policy, and leadership, a position that she shared with only four others among the 32 participants in the forum. This situation again demonstrated the principle of the noisy outlier. Mouinette peppered her contributions with shouted words (that is, typed in upper case letters), glowing compliments of the United States ending in exclamation marks, and derisive comments about Europe. This was written in the French of a native speaker, it seemed, although she indicated the US as her place of residence. Four other participants expressed support of the US, but three of these only contributed once to the discussion. In contrast, 17 participants critiqued the US on various grounds or pointed out signs of American vulnerability.

Typical of Mouinette's comments is:

> When China replaces the US (if it happens it won't be in 10 years) you will regret the loss. Why so much emotion in France around the AZF explosion [a terrorist attack on a chemical factory in Toulouse, September 2001] which caused an absolutely NEGLIGIBLE number of deaths relative to people who die on the road??? Same thing for Madrid [terrorist attack on Spanish trains, May 2004]: 400 deaths, that's nothing at all!
>
> (Mouinette moqueuse, October 20, 2004)

Here we also are dealing with sarcasm and her screen name in this instance, "mocking Mouinette" indicates as much. The main point we can gather is that French public discourses, like those in the US, tend to exaggerate the threat of terrorism and underreport more common everyday threats, but what makes Mouinette "shout" is that she assumes her view is the minority view on this forum. The importance of her comments for our analysis does not lie in the self-evident message content but rather in the tone as a kind of litmus test for what kinds of ideas seem so marginal they must be shouted (capitalized and set off in exclamation marks) for the speaker to believe he or she has been heard.

Removing Mouinette's contributions from this thread reveals a general consensus among the "quieter" voices: the number of comments critical of the US then outnumbers pro-US comments by a ratio more than 5 to 1. This consensus emerged against a double backdrop of angry comments by Mouinette and spurious remarks that had more to do with other participants than with the US election. In fact, forty-eight percent of the postings in this discussion consisted of tangential comments, for example a volley of remarks about China's treatment of Tibet, as well as speculative remarks relating to the identity and mental competence of other participants in the forum. Removing the "noise" of spurious comments and the "noisy outlier" of Mouinette reveals a geopolitical motif – the image of the US as dangerously strong and crucial points of weakness that could lead to its failure and a catastrophic collapse. This is, of course, a mixture of *Hyperpuissance* and the *Tottering Giant* motif (see chapter 4). The flavor of these critical comments is captured by three postings that can be lifted out of the context of the discussion and remain largely comprehensible.

> Dear Mouinette (moqueuse?!) I think that what cenoxon [sic] wanted to say is that for 3,000 deaths the Americans did not hesitate to go to war against a country that had nothing to do with the famous 3,000 deaths, thereby adding 1,087 (soldiers killed, not counting the seriously injured handicapped for life…) A war that caused through bombardment hundreds of thousands of civilian casualties… While the American state protects the bearers of weapons that cause 12,000 deaths a year. For the deaths of AZF or Madrid, there was no such stupid vengeance. Compare what is comparable. In regard to China, all that is necessary is that tomorrow, for reason X, they decide to sell part or all of the US treasury bonds they hold, then the dollar is a dead loss! In effect the American economy is in the hands of Far East markets: China, and also India and Japan.
>
> (nonna, October 25, 2004).

The second posting in this series adopted the distinctive style of an endless run-on sentence without capitals (which I have done my best to render in English).

> Best to stop closing our eyes in every way everyone knows bush waited for this in order to attack iraq and his goal was not to defend the united states but above all to gain control of petroleum just one question in your opinion with all the means the united states possessed how is it possible they have not located bin laden??? i'l simply say that everything is false the united states manipulates their citizens and us too reflect seriously on it …
>
> (romane Nov. 11, 2004)

And finally there was a warning about the economic vulnerability of the US, in a more conventional writing style.

> The federal budget deficit [of the US] is greater than that of France, it's approaching 5% but it's not the most important thing because as you argue it can be controlled. What is more disturbing is the trade deficit, 54 billion for the month of September. I remind you that only a few years ago the announcement of a billion per month made the markets collapse on Wall Street, today we're at more than a billion and a half per day and the figure is taking off…These cumulative deficits and the blatant autism of the United-States government represent the most serious dangers facing the planet. We [the French] don't know it, but the Iraqis become conscious of it a bit more each day. Best Wishes.
>
> (Agitprop, October 28, 2004)

Each of these contributions reveals a paradoxical connection between American strength and weakness and the disturbing knowledge that when a giant falls it invariably does damage to those unfortunates who are caught underneath. In these comments, the US is both aggressive and inward-turned, "autistic" to use the terminology of Agitprop (a term transplanted from governmental and journalistic critiques of George Bush, Chapter 5). Weakness and aggression go hand in hand, because the fundamental (economic) weakness of the US is masked from American citizens through a pointless and harmful show of military strength. Two of the three contributors quoted above were one-time participants in this discussion, while the last, Agitprop, contributed ten times. The convergence among views of one-time participants – generally critical of the US and of Bush – is particularly striking because it suggests a vast body of "lurkers" in this discussion forum who contribute only once or not at all but who are generally sympathetic to the *Tottering Giant* perspective. Conversely, the view of the US as a "friendly giant," that is, a defender of freedom without weaknesses, either moral or social, is a minority view in the online forum and therefore one that proponents "shout" with the use of capitals, sarcasm, and exclamation marks, generally creating far more postings than people who feel themselves to be among the majority.

We can conclude from this encounter with discussion forums that the Internet, unlike mass media, gave free rein to strong opinions about the United States. Furthermore, it was an uncontrolled space, where "balance" was provided not by editors upholding an ideal of balanced and impartial reporting, but rather by a struggle between unequally-matched opponents. In its agonistic way, the Internet also provided a kind of balance, but it was a balance between a relatively silent majority of one-time contributors and lurkers versus a vociferous minority. If this communication situation has something special to offer to politics, it is, first of all, the opportunity for people to be engaged participants rather than passive or at best indirect participants in political discussion. To "speak" (write) one's viewpoint is to open oneself up to the views of the other, even if it seems that none of the other participants are listening. By removing the inhibitions that are present in face-to-face communication, each side of the political debate has an opportunity to present itself; everyone "present" (those reading the discussion which is far more than the list of participants) can judge who makes the most convincing argument. Perhaps it is ironically the "lurkers" on the Internet that make it different than mass media, because their *possibility* of contributing shifts the way in which they "listen."

Secondly, this network of political discussion crosses national borders in a way that is at once pervasive and elusive. A conversation may include Americans of local descent, French expatriates living in the US, Canadians, French citizens, and conceivably anyone else with access to a computer and a rudimentary knowledge of the French language. The conversation spills across national borders and fluctuates through time as one group of participants comes home from work, converses, goes to bed, then makes way for another group of participants to take up the thread of conversation. This is not an idyllic discussion in which nationalism is transcended, but rather a confrontation of various nationalisms in a space that invites them to collide. The virtual place of the Internet forum permits not just anonymity but also geographical ambiguity. Participants in the debate want to remain placeless so they can speak with the French as French, even if they are Americans or Canadians.

Blogging the election

A blog consists of an online journal and comments from readers. The term "weblog" was coined in 1997 by Jorn Barger, and shortened to "blog" in 1999 (Blood 2000). Bloggers originally functioned somewhat like Internet tour guides, discovering online news and websites and providing readers with both links to, and comments (frequently ironic) about other sites. In comparison to the online forums just considered, blogs have a more personal character and generally preserve a greater level of civility or "netiquette."

Blogs can be as conflictual as discussion forums like the Wanadoo forum, but often they attract a following that is basically in agreement, so the discussion is less contentious. This was the case with the French blogs that covered the US election. These blogs were associated with French media (three newspapers and a television station) a factor that narrowed the range of opinions "voiced" online, making for somewhat tame discussions. The most prolific blog (in part due to the blogger and in part due to the participants) was maintained by Fabrice Rousselot, a foreign correspondent for the leftist newspaper, *Libération* (see chapter 5 for more on *Libération*). Rousselot's colleague, Pascal Riché, also had a blog hosted by *Libération* and he was the one who on March 26, 2004 posted anti-Kerry anti-France images copied from conservative American websites (see chapter 3).

Rousselot's post of September 28 generated seventeen responses over three days. This was a particularly interesting exchange that revealed the dynamics that drive online debate. The exchange is interesting with regard to what was not argued – no one advocated Bush – and with regard to what was argued: some saw Kerry as a potential source of change in the US-France relationship (*Big Difference* motif) while others chose to minimize the difference between Bush and Kerry (*No Difference* motif). Most of this exchange, which occurred just before and during the first debate between the presidential candidates, is reproduced here:

> So are you ready? General lunacy here waiting for the historic debate. Alan Schroeder, professor at Northeastern University and expert in such things, tells me this should be a good one. I'm dreaming of finally seeing these two punch it out face to face since they've battled each other from afar at each step of this campaign. Joking aside, it's a crucial moment for Kerry. If he manages to nail Bush on Iraq [*enfoncer le clou de l'Irak dans la botte de Bush*] anything could happen. And yes I have said that this campaign would hinge on Iraq and I did bet that Kerry will win with less than a five point lead.
>
> (Rousselot)

By this point in the campaign, Rousselot had attended the Democratic and the Republican conventions and attested to being verbally attacked by delegates at the Republican convention. His preference for the Democratic party was clear and his prediction of a Kerry win was therefore an optimistic one. Despite *Libération*'s leftist politics, support of Kerry by the newspaper's readers and reporters was not automatic, as the first two replies to his posting indicate.

> I think so too (I've even bet a bottle of Chateau Lafitte). I think the manipulation of the system goes clear to altering the results of opinion polls. People manipulate voting machines, why not opinion polls? My American friends tell me that Kerry is not much better. There's as much difference as between Pepsi and Coke. They're just the representatives of two lobbies (indeed the same one). Your opinion?
>
> (Laurent)[1]

> Mainly in agreement with you [Rousselot]. The French cry "Kerry! Kerry!" while jumping around like goats... But I'm persuaded he'll follow the same policy as preachy Bush, maybe with a bit of Vaseline...
>
> (marabbeh)

These posts minimizing the differences between the candidates most likely reflect the past century of US-France relations during which the French have most often perceived the Americans as acting out of self-interest regardless of the political climate in the US (see chapters 3 and 4). They also reveal a perspective to the Left of the entire American political spectrum that therefore minimizes the difference

1 This contributor included his last name which has been omitted here to preserve anonymity.

between Democrats and Republicans. This is the perspective that *Libération* could easily have taken and did not (chapter 5).

The next posting consisted primarily of a link to a copy of an Op-Ed article from the New York Times by Daniel Ellsberg, demonstrating the intertextuality of blogs. The next two posts expressed support for Kerry, although their argumentation tactics are hardly convincing.

> Ah, finally someone who refuses to retract their predictions. Thanks Mr. Rousselot. Yes, Kerry will win. And why not? Let's not prepare ourselves for his fall, despite the crushing majority opinion on the media… What does this mean? It only means there was a gigantic campaign (including in France) against Kerry. Nothing else. If Kerry wins, the world will win, and we will win… Do we have any interest in losing? No. And the Americans, are they ready to lose? No. Conclusion: in order for Bush to win the results at the voting booth will have to be incredibly manipulated. I bet also that Kerry will win with a several point margin."
>
> (Fernand Lechien)

> "as much difference as between Pepsi and Coke" And who designated a chain of countries as the "axis of evil"? Who preaches that we battle them one by one on a pretext of a "war on terrorism"? Oh, maybe there's a little difference, eh? A serious difference!
>
> (Méléis)

The intensity of the debate between those who saw a difference between the two US presidential candidates and those who did not suggests a dynamic that is not typically found in discourses about a foreign political campaign. Instead of being "out there," the American campaign had become provocatively part of the here and now in France. Since the outcome might have grave implications – since it might matter to the French who the Americans put into the White House – the *No Difference/Big Difference* debate was of more than passing interest to French observers.

The presence of this debate reflected the dilemma of a democratic society living in the shadow of a hyperpower. The awareness of France's democratic deficit generated both movements – towards a kind of virtual involvement by supporting Kerry from afar, and away from involvement by reasserting that the distant political contest would make little difference. The next three posts dealt with coverage of the presidential debate in France and with the tightly scripted nature of the debates, again adding to intertextuality. Following this was a long post from a French expatriate living in the US, who impressed subsequent participants with his insight:

> A fair number of people think that the debate itself won't be so decisive… but that in contrast the post-debate will determine the winner. In 2000, just after the Bush-Gore debate, surveyors gave a 14 point lead to Gore. For 2004 the Democratic commentators who talk just after the debate won't have as much weight as the conservatives (Giuliani and McCain) and the latter won't cease to hammer in with eloquence that Bush was the winner of the debate… [ellipses in original] leader… clear positions… war against terrorism… and that Kerry was not convincing… flip-flopper… weak… The cable channels in the United States are strongly conservative… they demonstrate it shamelessly and without

respite. Le Figaro, L'Express [newspapers] on the right? Here [in the US] they would be models of neutrality. These channels don't miss the chance to reflect constantly on the worst moments for Kerry (sweat…, hesitation…) and will take for fact the opinions of the commentators mentioned above. Unfortunately this is what people will remember. …

Therefore, if Kerry is to draw some kind of benefit from this debate he must perform exceptionally. And even if he does, I doubt the benefit will be significant. The subject of the debate Thursday will be the international situation (Iraq, terrorism), precisely the subject where Bush excels in the opinion polls. Bush's position is hard to refute because he has taken the moral stand that the war in Iraq is a matter of liberty and democracy versus oppression and what is needed is time for the country to stabilize, then the jackpot: stability in the Middle East, end of terrorism, in short, a "wonderful world." The amalgamation of the war against terrorism and the occupation of Iraq has taken hold. That's the source of my pessimism. I just hope the debate allows Kerry to address the disastrous flip-flopper image his detractors have given him. [Hypertext links to sites of the New York Times, Washington Post, and two American blogs are included.]

(Pierre)

The next post, from another expatriate (not included here), commended the "Beautiful analysis from Pierre," but optimistically maintained that Kerry would win. It is rare in online debate to find such positive inter-participant comments, but their presence here indicates the potential of this space to support the development of political consensus. The following post echoed these positive sentiments and added intertextual references to film and television, indicating that the Internet does not just complement these other media, rather it helps draw them into Democratic civil society.

Very nice analysis from Pierre :) Propaganda is at its height. I think nonetheless that this lie is so big it will end by "exploding" in the face of GB and his pals. I'm thinking of the lie about the absence of American deaths in Iraq, maintained by the lack of images. There have been enough images of Americans getting their throats slit to reveal this lie with a rare violence. This much said, it may take some time. In passing, I can only congratulate Mr. Moor [sic] who has been criticized by certain kind souls for the "trivial" side of his attacks. When you're in a war, you have to be ready to get your sleeves dirty. In addition I'd say that if I were an American I'd vote for Kerry while remaining cautious and aware that he's also at the mercy of the system. Let's hope there's enough disgust to provoke (at least among the enlightened class) a desire for change. In any case, that's what I believe.

(Laurent)

On this cue, the next participant expanded on the intertextual quality of meaning construction, drawing a parallel with mediated sports experiences.

It's strange this excitement and this pressure, it recalls the long hours of July 1998. I know, I know, drawing a parallel with soccer might seem stupid or incongruous, but taking a step back, we're talking about two events which attract quite a few people to the small screen. And quite a few people ready to yell 'Go Blue/Go Red/Go Yellow.' And if the intensity of

> the risks and the antagonisms were after all a drug? As an 'instant partisan' we [are able to] live deep and dense where the tension is clearly palpable here and elsewhere. This story pleases me so much that I've decided to leave for the United States next week to cross the country during the month of October and remain fifteen days after the election to see 'live' this strange moment in the 'democratic history of nations.' [Hypertext link to webcast of debate provided] Good night and good debate.
>
> (greg)[2]

This contributor is so intrigued by the election that he has decided to see it first hand. Mobility and communication become interrelated aspects of engagement with the historical event. More abstractly, travel and communication through films such as Fahrenheit 9–11 and live intercontinental television are frameworks in which the event *becomes* historical. This participant not only announces his imminent departure, but also provides another kind of voyage for his audience by adding the link to his page so that others can watch the webcast.

The last contribution to this exchange is very intertextual in character: "Thursday night's debate will be broadcast on France 2 at midnight Friday. I hope that the release of Moore's DVD (slated for October 5 on amazon.com) will nail Bush a bit more" (Babel). The intertextual links in this short note are fascinating: a film, released on DVD, is advertised on a website and that advertisement is, in turn, announced on the Blog. Through such intertextual links various foreign audiences are able more than ever to participate vicariously in the American political process: by sharing media, sharing information about media, and sharing reactions to the intertextual constructions of reality. All that is missing is a direct participation in the vote, yet discursive construction of the election in France works to build a kind of tentative and provisional consensus. This, in turn, can shape electoral politics in France, lending support to French candidates who capture the essence of political sentiments toward the United States. The "hidden transcripts" of resistance are not confined to hedgerows and saloons as they used to be (Scott 1990), but rather circulate through heterogeneous networks that replace the familiar geography of nations with a new geography of link topographies and areas of interest (Adams 1996). These mediascapes (Appadurai 1996) link between local and global scales, and disregard national boundaries.

Despite this flowering of physical and virtual space, reality still seems self-evident to at least one participant in this blog.

> Wow!!! It's getting wild on the TV and on the blog!!! I agree with Pierre on the influence of ideological discourse on the vote. One has the impression that many Americans dream of an America like the one Bush describes. How can they be drawn out of this dream of omnipotence and attain reality? The Europeans and the French, with a discussion closer to reality cannot make themselves understood, and vice versa.
>
> (jm, September 30)

2 This contributor included his last name which has been omitted here to preserve anonymity.

The theme here is *Continental Drift*, identified in the previous chapter. This participant continues with comments that set US, French and European culture in a three-way relationship and hint at the idea that *Europe Puissance* might offer a means of opposing US expansionism:

> Kerry won't be able to draw them out of their dream vision of America, but we can at least hope for an end to the unbelievable behavior of the Bush administration which consists of systematically increasing the tension until there is a confrontation, denunciation (*that* part is French, isn't it), threats and a power drive (like in basketball). However, like you see in matters of commerce, it's far from really paying off. What's more, in this kind of confrontation, you can notice that the Europeans make a united front, if only for their common economic interest. It must be really hard to work and make plans with the people in this administration.
>
> (jm, September 30)

Here we see indications of an intuitive understanding of the hard power/soft power contrast. The earnestness of this posting contrasts with the irony in the next posting, but both are, in a sense advocating disentanglement from the US.

> A few hours before the debate, the Iraqi terrorists say "peek a boo!" by blowing up several cars… and Schwarzie shows the way to his pal Bush: he's prohibiting the forced feeding of ducks and geese… Shi-(!)…from now until 2012! It's screwy!
>
> (gimik, September 30)

The exchange continues. There is speculation about the veracity and significance of opinion polls and advice on how Kerry could sway the undecided voters. Again, alternation between advocacy (of Kerry) and ironic detachment from the election defines the character of the forum. *No Difference* versus *Big Difference* is a lively debate among these participants. Insofar as participation in this Blog probably indicates interest in *Libération* and identification with the French Left, we can surmise that the Left had appropriated from scholarly discourse these motifs and the tottering giant motif to frame its participation in a deterritorializing world order.

On the whole, the blog served as a context for discussion among like-minded observers, which is not to say there was nothing to discuss. But here the discussion served as much to generate a shared viewpoint as a context for dispute and disagreement. Whether cynical about the US or fascinated by it, all of the participants seem to agree that the Bush administration presents a major problem. The blogs indicate some form of consensus more often than the discussion forums, perhaps because of the structuring role of the blogger.

Conclusion

For the French, the Internet provided a virtual gathering space in which to discuss and debate the 2004 American presidential election. This debate mobilized various themes of US-France relations, some old and others new, most drawn from other media via intertextual connection. The richness of intertextual links is a primary asset of the Internet as a communication medium. Surprisingly, despite the un-intellectual character of many of the postings, the Internet was more likely than television or newspapers to pick up on themes raised in academic writing. The US was seen rather ambivalently as overwhelmingly powerful and seriously vulnerable, in the motifs *Hyperpuissance* and *Tottering Giant*. When these images were evoked another giant came to mind – China – and insofar as China has more people and a healthy export economy, it appeared to have a more promising future than the US. The US was not only incurring a massive debt, but it was violently aggressive and inward turned at the same time, "autistic" in the recurring term, and its mass media (excluding Moore's "Fahrenheit 9–11") were seen as contributing to the problem. In regard to the two candidates, the clear consensus (despite a few noisy detractors) was that George Bush's policies were completely opposed to the values of the French people and French civilization. But debate raged on whether Kerry would be any better than Bush, particularly on the Left. One of the most intriguing motifs is the idea of a stronger Europe that would presumably emerge out of the framework of cooperation within the European Union as fostered by the Bush administration's intransigent stance.

Relative to other communication media, the Internet holds a unique position in civil society. Its range of communication contexts includes one-to-one, one-to-many, and many-to-many topologies. In the forums and blogs considered here, it is common to address a particular individual by screen name. This is superficially one-to-one communication, but the other participants "listen in" and add comments supporting one or the other participant, and sometimes they mock one or the other of the participants or try to determine the location of a participant all for the benefit of other participants. A discussion forum is therefore a many-to-many communication situation while a blog is many-to-many with one person playing an agenda-setting and gate-keeping role.

The prospect for reasoned political debate via Internet discussion forum or blog sometimes appears rather bleak because of the high ratio of personal, irrelevant, and incoherent comments. The common phenomenon of the "noisy outlier" may seem to exacerbate the shortcomings of the Internet as a public sphere, but the provocation of such gadflies generates many more responses than if participants were largely in agreement. The intertextuality of Internet also appears on first consideration to erode political discourse because so often participants include only a link without any personal contributions, and the media they are linking to may be commercial media. But the ability to "borrow" from other web sites and from other media such as television, newspapers and magazines complements the strengths and weaknesses of the hastily typed postings that form the basis of Internet forums and blogs. This linking provides logical context and evidence supporting the viewpoints of participants and grounding the discussion in formal debate, academic commentary, online news and other sources of verifiable information, all the while permitting commentary on these other media.

As it currently stands, the Internet has some flaws as a medium of democratic discourse, but these flaws are very different from the flaws of television and newspapers: polyvocality versus univocality, chaos versus rigid order, and anarchy versus top-down control. Intertextuality in particular permits the Internet to serve as a valuable complement to these other media, especially in light of its affordances of interactivity, instantaneity, anonymity, transnationality, and topological multiplicity. The ambiguity of Internet communication, a product of haste and dependency on typed text (which filters out many non-verbal aspects of speech) is both an asset and a liability for political communications as it stimulates participation but stands in the way of consensus-building. The Internet is therefore both exciting and frustrating from a progressive political viewpoint that actively seeks and supports the deterritorialization of the public sphere. In facilitating intertextual, participatory, transborder debate, however, it offers something that no other medium can offer at this time.

Chapter 8

Quel Rapprochement?

> In politics shared hatreds are almost always the basis of friendships.
>
> Alexis de Tocqueville

How can we summarize these communications and at the same time clarify the conditions producing the Achilles heel of the hyperpower? If globalization proceeds via particular cases of two or more countries "coming closer," then in French terms we can ask of the 21st century: "quel rapprochement?" What countries will be coming together and why? Certainly media of various kinds including television, newspaper, Internet and academic writing will all be involved, each in its own way. Certainly powerful countries like the US and France will be involved. But we cannot "read" Americanization off the expansion of any one medium, even if a count of publishing houses or Internet servers indicates a numerical dominance by the United States. The geopolitical impact of globalizing communication is more complicated than that.

As long as the US is the most powerful country in the world its military actions are likely to be seen as self-serving and its role as global policeman is likely to arouse suspicions in international discourses. Perhaps the most Americans can hope for in foreign eyes while their country remains as powerful as it is, is awareness of the US's internal diversity – its numerous regional and social divides, the particularity of its places, the humanity of its people. This kind of geographically nuanced perception of the US defuses blanket condemnations of Americans. Internet discourses of fall 2004 tended to generalize about America and Americans whereas the mass media as indicated by television and newspapers avoided that trap. On this account the French mass media were not anti-American although they did expose an alien (to the French) part of the US. In contrast, the Internet was intriguingly instrumental in developing a pro-European view and drawing on themes in scholarly debate despite its chaotic discussion style.

I have shown that French media and Internet communications are in the process of producing a particular geopolitical perspective on the US through the interconnection of various motifs. Some of these motifs, including *Divided America, Tough Campaign,* and *Big Difference*, emphasize the internal division of the US. These motifs help defuse anti-Americanism by invalidating general comments about America or Americans. Their presence indicates the French media in a role that is neutral to supportive of the US because they leave room in the American landscape and society for a change for "the better" from a French point of view. Other motifs, however, construct the US as more of a threat: *Tottering Giant, No Difference,* and *Continental Drift* together paint a picture of an America that is growing away from Europe in terms of its core values, becoming violent and unstable at home and abroad, while its electoral process is antiquated and flawed. The fact that the latter set of motifs was less prevalent than the former indicates that claims of anti-American media bias in France are unfounded.

The tension between various groups of motifs indicates, as well, the linkage between geopolitical and topographic scales, or stated more simply, the global and the local. Voter profiles (in the case of newspaper) and in-depth reporting on small towns (in the case of newspaper and television) presented the conservative pole of the US's political divide in great detail, lending insight into certain Americans and a certain America. Bush's popularity was given meaning within this symbolic framework by the forces of geographical isolation of at least two distinct kinds acting on voters. American politics, and hence the US's role in world affairs, became an epiphenomenon of American rurality arising from the vast spaces of the West and byways of the South, and the brutal racial and economic segregation of American cities. It made sense that American voters, isolated from the world by the peculiarly isolating American geography – both urban and rural – responded to the crisis of September 11, 2001 by turning toward nationalism and a theocratic value system. The real question was whether this parochialism would gain the upper hand in 2004 and thereafter. The debate in France over the American election therefore boiled down to a debate about American geography, politics, and society, not about which candidate would be the better leader.

Turning to particular media, we can elaborate on affordance-based political functions. The incoherence of the Internet, its peripheral comments, and its garrulous "outliers," opened up the agenda of the political debate. The Internet appears to have generated little genuine understanding of the US but great involvement from participants in France and beyond. The participants in this electronic public sphere could freely express their opinions, however conventional or bizarre, however reasonable or incoherent, however sympathetic or cynical. Their open-ended conversation in turn made the Internet part of civil society in a way that media like television and newspapers are not. This was a far cry from the passive role envisioned by an American critic who wrote in the *American Spectator* that "Despite their well-earned reputation for cynicism, the French usually swallow uncritically whatever line their government/media Establishment feeds them" (Harriss 2004–2005, 64). Instead, the online discussion critically engaged with the French and American media, weaving together participant views and references to various other media, American and French.

If the French obtained a sense of the particular places in which American political attitudes had developed, it was most likely through either the newspaper or television, or both. These "mass" media together created a strong sense of the localization of political sentiments in the US. As centralizing and naturalizing media, and as media that define the past and present, these media therefore offered a powerful antidote to anti-Americanism with its homogenizing gaze on America and Americans. This function, unappreciated from the other side of the Atlantic, was in direct contradiction to American critiques of French media as the "faithful servant of the Élysée Palace and Quai d'Orsay" (Harriss 2004–2005, 64). The Internet may have fueled anti-Americanism as a component of democratic debate, but quantitative analysis of over 350 discussion forum postings indicated that the volume of comments critical of France and the French government almost perfectly matched the volume of comments critical of Americans and the US government. The participants paid less attention to the EU than to the US in the discussions considered here, but the EU was strongly supported among those who mentioned it.

In essence, by representing politics (American politics) the French were in fact *participating in politics* although in a politics of a different order – a deterritorialized, boundless communicative politics of citizens living at once both inside and outside of the state where they reside. If the politics in the public eye, that is, the American election, was a brief episode, the politics forming across borders was of a slower and more gradual nature, but no less important. This politics is not devoid of electoral impact, but the impact is on French and European governance and only indirectly on American government. As the US gradually loses control of its hegemony owing to its inability to completely control the economic forces it has set loose (Agnew 2005) this unbounded kind of political representation (Low 1997) will increasingly shape the emerging 21st century world order.

Nothing conveys this fluid situation as well as the Internet debates in which participants in North America debated with participants in Europe about the upcoming American election. It has previously been noted that political struggles do not simply occupy the spaces in which they occur; they actually create spaces and contest definitions of space and place (see Adams 1995; Staeheli 1994; Mitchell 1996). Every political struggle in some sense involves contestation over scale, space, place, and virtual places of communication. As their discussions drew in references to presidential speeches and debates, as blogs featured copies of "France-bashing" bumper stickers, and as observers in France, the US and elsewhere "borrowed" jokes and other material in English to post on French websites, a virtual gathering place and a means of transforming political scale were crafted. The Internet discussion intertextually drew on and contributed to a shared virtual space of the media to which all participants had some degree of access. The space in which this ephemeral political formation existed – the *bridgespace* (Adams and Ghose 2003) spanning the Atlantic – was accessible from both shores. Each Internet discussion, scholarly essay, newspaper article or television report constituted a particular place of political engagement within the larger transatlantic bridgespace. The purpose of such deterritorialized political forums will ultimately be the maintenance of peace, as place-transcending communication has provided, and will continue to provide, a necessary complement to the bounded politics of the polling booth. It will do so by mustering political sentiment in an ad-hoc way to meet antidemocratic forces arising at the global scale from organizations and from democratic and non-democratic governments.

It is clear that actions taken in violation of international law and/or unilaterally by the United States become liabilities to US soft power through the process of reverberation because they prompt criticism from citizens outside the US. International criticism, resonating between intellectuals of statecraft, media, and "ordinary" citizens around the world can affect subsequent political decisions outside the US, as voters in other states sanction their leaders for cooperating with the American government, or as politicians in other states take positions opposing the US in order to garner favor among the voters. So if American citizens do not act as a check on the power of their own government, then other citizens will do so indirectly. To see this all that is required is time for citizen preferences to affect politics in their respective republics and for those republics to develop cooperative means of exerting power to constrain a source of power that is no longer internally constrained. The Westphalian system has broken down, in part because of the concentration of power in a single hyperpower, the United States, and because of questionable decisions made by American leaders, but also in part because of reverberation of political thought – of representations and their constituent motifs – through various modern communication media at the level of the "ordinary" transnational citizen.

Manuel Castells predicts in *The Information Age: Economy, Society & Culture* that "Nation-states will survive, but not so their sovereignty. They will band together in multilateral networks, with a variable geometry of commitments, responsibilities, alliances, and subordinations" (1998, 355). He goes on to note that the European Union will be the "most notable" of these multilateral networks. The emergence of the transnational citizen through communications of diverse sorts will lead the way for the re-emergence of European power at the global scale, but its citizens probably will continue to envision themselves in cosmopolitan or "cosmopolitical" (Cheah and Robbins 1998) ways. Meanwhile, Americans may unfortunately continue to constitute their perceptions, conceptions and representations of the world at the national level, nursing a worldview that is embattled by former allies on the discursive level and by enemy states and networks on the physical, material level. The source of the challenge is ultimately power itself – the US's own unsustainable hard power and the media environment that causes that power to become a liability. American supremacy in regard to economic and military power and the antagonism of citizens to cultural and political hegemony virtually assure this state of affairs. Meanwhile, defensive views inevitably circulate within the overwhelmingly powerful nation in the form of aggressive and/or self-congratulatory representations which lend power to conservative political factions, and subsequently spill over national borders to cause offense on the part of other states. As citizens of weaker states respond to the democratic deficit induced by the existence of a hyperpower and its internal and external dynamics their critique is stripped of content by American observers and translated into anti-Americanism, a nationalist reaction that further exacerbates American vulnerability.

Europe and Peace-Making

Europe's position in this emergent internationalism is particularly interesting. Kant's "Perpetual Peace" was most likely inspired by the signing of the Treaty of Basel

(Wood 1998, 59) although this treaty would prove no more effective than the Treaty of Westphalia at ensuring peace in Europe. Nonetheless, Europe's hard-won state of peace bore fruit in the form of Kantian internationalism. In his landmark formulation of orderly inter-state relations, Kant made a case for the republic, as opposed to the monarchy, and for the implementation of a system of international law among republics. What is of inestimable importance with regard to the current world order is that the basis of Kant's preference of the republic over the monarchical state lay in the premise that power corrupts. Limitation of power through the mechanisms of representational politics creates a state that is therefore less corrupt and is capable of pursuing the common good of its citizens, which entails among other things the avoidance of war.

> Now, in point of fact, the Republican Constitution, in addition to the purity of its origin as arising from the original source of the conception of Right, includes also the prospect of realising the desired object: Perpetual Peace among the nations. And the reason of this may be stated as follows.—According to the Republican Constitution, the consent of the citizens as members of the State is required to determine at any time the question, 'Whether there shall be war or not?' Hence, nothing is more natural than that they should be very loth to enter upon so undesirable an undertaking; for in decreeing it they would necessarily be resolving to bring upon themselves all the horrors of War. And, in their case, this implies such consequences as these: to have to fight in their own persons; to supply the costs of the war out of their own property; to have sorrowfully to repair the devastation which it leaves behind; and, as a crowning evil, to have to take upon themselves at the end a burden of debt which will go on embittering peace itself, and which it will be impossible ever to pay off on account of the constant threatening of further impending wars.
>
> (Kant 1983 [1795])

As "the best known piece of writing by any major philosopher on the topic of peace between nations" (Wood 1998, 62) Kant's essay has been subject to various critiques, both from "realists" such as Robert Kagan (2003), who assert the primacy of force rather than law in maintaining order between nations, and from post-colonial theorists such as Pheng Cheah (1998) who argue that capitalism undermines any global federation by concentrating power. Consideration of television, newspaper and Internet communications as well as academic debates, as they reverberate with US politics from outside, offers a fundamentally new type of insight on the question of Kantianism in the current world order.

The conundrum posed by the existence of a hyperpower is that it can engage in war without imposing a significant cost on ordinary citizens. The wars in Afghanistan and Iraq serve as cases in point. The capitalist tendency for wealth, and therefore power, to accumulate and concentrate indicates a limit to Kant's interstate system. If power corrupts politics within a state, as Kant argues, it can certainly do so between states. As one state dominates all of the others Kant's logic no longer applies. The existence of a hyperpower threatens the peace between republics even if that hyperpower is a republic and even if its ideologies include the heroic ideas of promoting democracy and shouldering the burden of world peace. Insofar as peace leads to prosperity,

prosperity depends on trade, trade produces wealth, and the seeds of war are sown in the concentration of wealth which permits one republic to incite war at little cost relative to its total wealth, population and security. It makes little sense to fault the US for having achieved global dominance, but its position is not stable, due to the democratic deficit sensed by citizens of other, less powerful (wealthy) states, and the subsequent breakdown in the Kantian cost-benefit logic of war avoidance. The study of geopolitical discourses therefore reveals the fault lines and points of growing tension in the world system.

An unbounded kind of democracy is emerging that depends on diverse media and telecommunications. News coverage of foreign elections through the mass media is one element of such boundary-crossing political communications, although it does not necessarily entail international discussion in and of itself. The Internet alone is scarcely more useful, despite its rather remarkable affordances. But when the Internet is combined with mass media via intertextual discussions and debates, so as to provide elements of interactivity and globalization as well as a solid grounding in rational thought and the particularities of place and personality, then the rudiments of trans-border civil society are present. Many of the communication flows in such hybrid communication spaces will not be flattering to the US, and there is no reason they should be. Their function is to act as a check on any threat to citizens in general, not simply to safeguard the interests of any particular state, including the US.

Their democracy-promoting role can be contrasted with that of official propaganda disseminated by the United States. The aim of the Bureau of Information, under the direction of the Bureau of Public Affairs is to "Inform and seek to influence foreign opinion-makers by presenting U.S. positions on policy issues through a variety of products" (Clinton 1999). This mission is in essence to promote propaganda through such means as the Voice of America, Radio Free Europe, Radio Free Iraq, Radio Free Afghanistan, Radio Free Asia, and other projects of the United States Information Agency and, since 1999, the Bureau of Public Diplomacy and Public Affairs. The "problem" of soft power from a strategic American viewpoint cannot be so easily fixed because emerging forms of rooted cosmopolitanism (Appiah 1998; Cohen 1992) are replacing nationalistic modes of citizenship and Westphalian realist politics regardless of American dreams or schemes. For ethical and practical reasons, media and telecommunications must be envisioned as a multi-directional reverberation rather than a one-way flow. The purpose in rethinking the directionality of international communication flow is to dismiss efforts to "fine tune" the American image abroad with propaganda and public relations, even from a viewpoint as nuanced as that of Joseph Nye.

As long as the US remains dominant in economic, political, and military spheres, and arguably in the sphere of culture as well, it will be consistently or sporadically opposed by those outside of US territory who are interested in strengthening democracy, as well as those who have no democratic intentions. The problem is hyperpower itself and the problem cannot be fixed by attacking formal or informal representations of the US any more than it can be fixed by attacking "rogue nations" and terrorist cells. The notion of the Achilles heel of the hyperpower implies that the benefits to be obtained by politicians through critiquing US power remain significant whether the leaders in question live in France, Britain, Egypt or Indonesia. Democracy facilitates the diffusion of this anti-hyperpower stance rather than discouraging it, because citizenship builds an expectation of sovereignty, media reveals the illusory quality of that sovereignty, and a perception of injustice arises among citizens or would-be citizens when external forces are seen to impinge on their nation's sovereignty.

The Way Forward

The EU serves as a harbinger of transnational or supra-national citizenship. Nye's seminal contribution to the recognition of this hazard is the recognition that force alone is not a basis for a strong position in the international arena. Nye's limitation is in thinking mainly in *strategic* terms even as he rightly turns the focus toward rhetoric and communication between states. The question is not how to buttress the US's soft power but how to step back from the temptation to try to be a hyperpower, even if drawn to the dream of being a "self-denying and benevolent leader" (Agnew 2005, 28) to recognize the anti-democratic impulse in this aspiration to "lead", and to permit democracy to flourish above the level of the state, and therefore at a scale of geographical aggregation where US power is potentially challenged. The ethically and morally justifiable goal is not for the US to attempt to interfere with political communications reverberating in the realm of international communication and to reassert its soft power – for example by planting pro-American participants on Internet discussion forums. This would simply give a more devious face to American nationalism and "leadership." The US-backed participants could vociferously back the US to little avail since the model of the noisy outlier shows that consensus forms among those who participate the least and is strongly linked intertextually to other media. The goal should be, instead, for scholars in the US and elsewhere to recognize and articulate the primacy of citizenship over the interests of any given state or nation, and to promote the growth of democratic discourses in the proliferating and accelerating global networks of communication. These networks are already transcending and will increasingly transcend the bounded spaces of the state. The persistence of democracy in the face of globalization will become more certain as: (a) citizenship breaks the bounds of the state and takes root in international communication networks, (b) civil society exploits the affordances of various media to the fullest extent rather than engaging in flaming and other irrational and oppressive forms of communication, and (c) people are seen and heard by others as citizens in general rather than just as "Americans" or "Frenchmen," or even "Europeans," (d) media confluence through the use of intertextual and hypertext links generates multiple sources for ideas and information.

A single global government would have all of the weaknesses of a single state. It is not to be hoped for as a source of future world peace and freedom. But multiple state governments bound together and held in check by transnational citizens is greatly to be desired. These citizens would enjoy various forms of virtual gathering and communicative action across borders as well as internationally protected rights to communicate, and would thereby offer a new level of security to all of the world's citizens. A Habermasian public sphere at the global scale could resurrect the elements of Kantian peace that are currently being erased by the emergence of a hyperpower. It would bring together diverse nationalities in a single space of citizenship. This would be a space between places – a bridgespace.

This rapprochement through the construction of bridgespace does not mean dominance or one-sided compromise. It is not an agonistic space of domination and resistance. Rather it entails mutual recognition by citizens, of other citizens, across permeable borders. These ephemeral shared perceptions, conceptions and representations in turn may support the formation of new political power rivaling the US, such as the EU or a core within the EU. And France is positioned to play a guiding role in this emerging entity.

References

Adams, P.C. (1992), "Television as Gathering Place," *Annals of the Association of American Geographers* 82:1, 117–35.

_____ (1995), "A Reconsideration of Personal Boundaries in Space-Time," *Annals of the Association of American Geographers*. 85:2, 267–85.

_____ (1996), "Protest and the Scale Politics of Telecommunications," *Political Geography* 15:5, 419–41.

_____ (1998), "Network Topologies and Virtual Place," *Annals of the Association of American Geographers*. 88:1, 88–106.

_____ (2004), "The September 11 Attacks as Viewed from Quebec: The Small-Nation Myth in Geopolitical Discourse," *Political Geography* 23:6, 765–95.

_____ (2005), *The Boundless Self: Communication in Physical and Virtual Spaces* (Syracuse, New York: Syracuse University Press).

_____ and Warf, B. (1997), "Introduction: Cyberspace and Geographical Space," *The Geographical Review*. 87:2, 139–45.

_____ and Ghose, R. (2003), "India.com: the Construction of a Space Between," *Progress in Human Geography* 27:4, 414–37.

Adams, P.C. et al. (eds.) (2001), *Textures of Place: Exploring Humanist Geographies* (Minneapolis: University of Minnesota Press).

Adrien, B. (2001), "Europe et Etats-Unis: tout est dans les têtes," *Commentaire* 94 (summer): 315–19.

Agence France Presse. (2005), "French Bashing Alive and Well in Parts of America," *yahoo.com* (published online 6 October 2005), <http://news.yahoo.com> accessed 17 October 2005.

Agnew, J. (1998), *Geopolitics: Re-visioning World Politics* (New York: Routledge).

_____ (2005), *Hegemony: The New Shape of Global Power* (Philadelphia, Pennsylvania: Temple University Press).

Albert, M. (ed.) (2005), *Regards croisés sur l'Europe* (Paris: Presses Universitaires de France, Académie des sciences morales et politiques).

Anderson, B. (1991), *Imagined Communities: Reflections on the Origin and Spread of Nationalism.* Revised edition (ed.) (London & NY: Verso).

Andreani, J. (2001), "Les Européens auront les Américains qu'ils méritent," *Commentaire* 94, 295–302.

Ang, I. (1985), *Watching Dallas: Soap opera and the Melodramatic Imagination* (London and New York: Methuen).

Appadurai, A. (1996), *Modernity at Large: Cultural Dimensions of Globalization* (Minneapolis: University of Minnesota Press).

Appiah, K.A. (1998), "Cosmopolitan Patriots," in Cheah and Robbins (eds.).

Aron, R. (2005), "Accueil," *Commentaire* <http://www.commentaire.fr> (home page) accessed 10 May 2005.

Artinian, P. (2004b), "Je voterai pour nader," *Le Monde*, 20 October 2004, p.5

_____ (2004a), "Surplus Chinois pour clients Mexicains," *Le Monde*, 19 October 2004, p.3.

_____ (2004c), "Cette fois, j'assure et je vote Kerry," *Le Monde*, 22 October 2004, p.5.

_____ (2004d), "Je suis à 100% 'pro-life': je vote Bush," *Le Monde*, 23 October 2004, p.3.

_____ (2004e), "Je vote Bush: Kerry a trahi son pays," *Le Monde*, 24–25 October 2004, p.4.

_____ (2004f), "Je suis bien sûr démocrate," *Le Monde*, 26 October 2004, p.5.

_____ (2004g), "L'économie ne marche plus, je vote Kerry," *Le Monde*, 27 October 2004, p.3.

_____ (2004h), "Je vote Bush, il défend mes valeurs," *Le Monde*, 28 October 2004, p.3.

_____ (2004i), "Face à la guerre, Bush est l'homme qu'il nous faut," *Le Monde*, 29 October 2004, p.7.

_____ (2004j), "À contrecoeur, j'irai voter Kerry," *Le Monde*, 30 October 2004, p.5.

_____ (2004k), "Tous les emplois s'en vont: je vote Kerry," *Le Monde*, 1 November 2004, p.3.

_____ (2004l), "Je suis indécis, je suis démocrate," *Le Monde*, 2 November 2004, p.3.

_____ (2004m), "Je n'ai jamais autant aimé Clinton," *Le Monde*, 3 November 2004, p.3.

_____ (2004n), "On a vraiment fait le mauvais choix," *Le Monde*, 5 November 2004, p.4.

_____ (2004o), "Nous sommes une grande nation," *Le Monde*, 6 November 2004, p.7.

_____ (2004p), "Bush a fait du bon boulot," *Le Monde*, 8 November 2004, p.3.

Assas Editions. (2005), "Assas Editions History," *Assas Editions* <http://perso.wanadoo.fr/assas-editions/assas-1.htm> (home page) accessed 10 May 2005.

Atkinson, D. (2000), "Geopolitical Imaginations in Modern Italy," in Dodds and Atkinson (eds.).

Balibar, É. (2005), *L'Europe, l'Amérique, la guerre: réflexions sur la médiation européenne* (Paris: La Découverte/Poche).

Barabási, A-L. (2003), *Linked: How Everything is Connected to Everything Else and What it Means for Business, Science, and Everyday Life* (Cambridge, Massachusetts: Plume/Penguin).

Barnes, T.J. and Duncan, J.S. (eds.) (1992), *Writing Worlds: Discourse, Text and Metaphor in the Representation of Landscape* (New York: Routledge).

Barochez, L. de. (2004), "Les Europeens on fait le pari de Kerry," *Le Figaro*, 26 October 2004, p.3.

Barthes, R. (1967), *Elements of Semiology,* translated by Annette Lavers and Colin Smith (New York: Hill and Wang).

_____ (1972), *Mythologies,* selected and translated by Annette Lavers (New York: Hill and Wang).

_____ (1974), *S/Z* (London: Cape).

Baudry, P. (2004), *Français et Américains: l'autre rive* (Paris: Village Mondial/ Pearson)

Baughman, J. (1987), *Henry R. Luce and the Rise of the American News Media* (Boston: Twayne Publishers/G.K. Hall & Co).

Beaudin, H. (2005), *Au-delà du « non »* (Paris: Ellipses).

Beaverstock, J.V. et al. (1999), "A Roster of World Cities," *Cities* 16:6, 445–458.

Berkowitz, D.A. (1997), *Social Meanings of News* (Thousand Oaks, California: Sage Publications).

Bernstein, R. (1990), *Fragile Glory: A Portrait of France and the French* (New York and London: Penguin/Plume).

Bertrand, C-J., (ed.) (1999), *Médias: introduction à la presse, la radio et la télévision.* 2nd Edition (Paris: Ellipses).

_____ (1999), "Les medias en France: situation actuelle," in C-J Bertrand (ed.).

Blonsky, M. (ed.) (1985), *On Signs* (Baltimore: Johns Hopkins University Press).

Blood, R. (2000), "Weblogs: a History and Perspective," (published online 7 September 2000) <http://www.rebeccablood.net/essays/weblog_history.html> accessed 15 August 2006.

Blyth, C. (2005), Personal conversation with linguist Carl Blyth of the University of Texas Department of French and Italian.

Bob-Morane. (2004), "Mon droit de vote américain," letter to the editor, *Libération*, 16–17 October 2004, p.7.

Boissonat, J. (2003), "Quelle place pour la France au XXIe siecle?" *Etudes* (May): 585–95.

Bollaert, B. et al. (2004), "Kaspi-Todd: une nouvelle Amérique ?" *Le Figaro* 27 September 2004, p.12.

Bollaert, B. (2001), "Adieu Union européenne, bonjour AELE?" *Commentaire* 93 (spring): 17–19.

Boniface, P. (2004), "Europe: l'élargissement au prix de l'affaiblissement?" *La revue internationale et stratégique* 54 (summer): 33–37.

Borneman, J. and N. Fowler. (1997), "Europeanization," *Annual Review of Anthropology* 26:487–514.

Bouygues (2006), The Bouygues Group <http://www.bouygues.fr/us/index.asp> (home page) accessed 17 February 2006.

Boyles, D. (2005), *Vile France: Fear, Duplicity, Cowardice and Cheese* (San Francisco: Encounter Books).

Brenner, M. and Parmentier, G. (2002), *Reconcilable Differences: U.S.-French Relations in the New Era* (Washington, D.C.: Brookings Institution Press).

Brenner, N. et al. (eds.) (2003), *State/Space: A Reader* (Malden, Massachusetts: Blackwell Publishing).

Buchanan, M. (2002), *Nexus: Small Worlds and the Groundbreaking Science of Networks* (New York: W.W. Norton and Company).

Caldwell, C. (2000), "Védrinism: France's Global Ambition," *Policy Review* 103 (October–November): 29–45.

_____ (2003), "Information Technology and the International Public Sphere," in Schuler (ed.).

Calvez, J-Y. (2004), "Pour rendre vivable l'univers asymmetrique," *Etudes* (Jan): 23–30.

Carroll, R. (1988), *Cultural Misunderstandings: The French-American Experience.* Translated by Carol Volk (Chicago and London: University of Chicago Press).

Casanova, J-C. (2004), "L'Amérique risque de s'enfermer dans une politique d'isolement," *Le Monde* 5 November 2004, p.6.

Caslon Analytics. (2005), "Le Monde and Le Temps," *ketupa.net Media Profile* ketupa.net a Caslon Analytics media resource (published online March 2005) <http://www.ketupa.net/lemonde.htm#temps> accessed 15 August 2006.

Castells, M. (1996), *The Rise of the Network Society* (Cambridge, MA; Oxford, UK: Blackwell).

Chambraud, C. (2004), "Les projets des deux candidats reflètent deux visions de l'Amerique," *Le Monde* 3 November 2004, p.4.

Cheah, P. (1998), "Given Culture: Rethinking Cosmopolitical Freedom in Transnationalism," in Cheah and Robbins (eds.).

_____ and B. Robbins, (ed.) 1998. *Cosmopolitics: Thinking and feeling beyond the nation* (Minneapolis: University of Minnesota Press).

CIA. (2006a), "European Union," *CIA World Factbook* <http://www.odci.gov/cia/publications/factbook/index.html> accessed 15 August 2006.

CIA. (2006b), "United States," *CIA World Factbook* <http://www.odci.gov/cia/publications/factbook/index.html> accessed 15 August 2006.

Climacus. (2004), "Que faire?" online forum posting at the website of *Le Monde*, www.lemonde.fr (November 3, 2004),

Clinton, W. (1999), "Presidential Decision Directive PDD 68: International Public Information (IPI)," (submitted to Congress 30 December 1998) <http://www.fas.org/irp/offdocs/pdd/pdd-68-dos.htm> accessed 15 August 2006.

Cohen, B. (2004), Bush et Kerry unis par la peur. *Libération* 7 October 2004, p.37.

Cohen, M. (1992), "Rooted Cosmopolitanism," *Dissent* (Fall 1992): 478–83.

Cohen-Tanugi, L. (1985), Le droit sans l'Etat.

_____ (1992), *L'Europe en danger* (Paris: Fayard, Librairie Arthème).

_____ (1995), *Le choix de l'Europe* (Paris: Fayard, Librairie Arthème.

_____ (2003), *Les sentinelles de la liberté: l'Europe et l'Amérique au seuil du XXIe siècle* (Paris: Odile Jacob).

Cohn-Bendit, D. (2005), "Relations franco-allemands et construction européenne," in Albert (ed.).

Colombani, J-M. (2000), *Les Infortunes de la République* (Paris: Bernard Grasset).

_____ (2002), Tous Américains? *Le Monde* après le 11 septembre 2001 (Paris: Librairie Arthème Fayard).

_____ and W. Wells. (2004), *Dangerous De-Liaisons: What's Really Behind the War Between France and the U.S.*, edited by L. Jacob-Duvernet (Hoboken, NJ: Melville House).

Conesa, P. (2004), "La vision américaine de la sécurité internationale," in Wieviorka (ed.).

Conte, C. (2004), "Européens et musulmans: parlons franc," *Panoramiques* 68:3, 124–32.

Coultrap, J. (1999), "From Parliamentarism to Pluralism: Models of Democracy and the European Union's 'Democratic Deficit'" *Journal of Theoretical Politics* 11:1, 107–135.

Cox, K. (ed.) (1997), *Spaces of Globalization: Reasserting the Power of the Local* (New York and London: The Guilford Press).

CREDOC (Centre de Recherche pour l'Étude et l'Observation des Conditions de Vie), (2003), "La diffusion des technologies de l'information dans la société française," a component of the survey: "Conditions de vie et aspirations des Français," Paris: Conseil Général des Technologies de l'Information/l'Autorité de Régulation des Télécommunications.

CSA. (2004a), "Les Français et l'election présidentielle américaine," *CSA* <http://www.csa-tmo.fr/dataset/data2004/opi20041028a.htm>

CSA. (2004b), "Les Français et l'election présidentielle américaine," *CSA* <http://www.csa-tmo.fr/dataset/data2004/opi20041103a.htm>

Dahlberg, L. (2001), "Computer-Mediated Communication and the Public Sphere: A Critical Analysis," *Journal of Computer Mediated Communication* 7:1 (online journal) <http://www.ascusc.org/jcmc>.

Darling-Wolf, F. (1997), "Framing the Breast Implant Controversy: A Feminist Critique," *Journal of Communication Inquiry*, 21:1, x–x.

Dassault. (2005), "Defense Activities," *Dassault Aviation* <http://www.dassault-aviation.com/defense/gb> accessed 15 August 2006.

Davis, J.K. (2003), *Reluctant Allies & Competitive Partners* (Washington, D.C.: Brassey's).

de Charentay, P. (2005), "L'Amérique en campagne" *Etudes* (September): 155–72.

de Montbrial, T., et al. (eds.) (1995), *Agir pour l'Europe: les relations franco-allemands dans l'après-guerre froid* (Paris: IFRI/Masson).

de Villepin, D. (2003), *Un autre monde ?* (Paris: Éditions de l'Herne).

Dean, J. (2004), "Les sombres mystères du clan Bush," *Le Figaro* 2–3 October 2004, p.12.

Debat, A. (2004), "John Edwards, co-équipier de John Kerry « Le people américain a soif de changement »" *Le Figaro* Magazine 2 October 2004, p.20–21.

Decker, F. (2002), "Governance beyond the Nation-State: Reflections on the democratic deficit of the European Union," *Journal of European Public Policy* 9:2, 256–72.

Delanty, G. (2002), "Communitarianism and Citizenship," in Isin and Turner (eds.).

Delors, J. (2003), "Une vision ambitieuse pour la Grande Europe," *Futuribles* 282 (January): 5–11.

Demangeon, A. (1920), extract from *Le déclin de l'Europe* (Paris: Payot) <http://hypo.ge-dip.etat-ge.ch/www/cliotexte/html/declin.europe.html> accessed 15 August 2006.

Dempsey, G.T. (2002), "One Less Follower: The U.S. dumps the ICC," *National Review* May 8, 2002.

Dijkink, G. (1996), *National Identity and Geopolitical Visions: Maps of Pride and Pain* (New York and London: Routledge).

_____ (1998), "Geopolitical Codes and Popular Representations," *GeoJournal* 46: 293–99.

Dodds, K. (1996), "The 1982 Falklands War and a Critical Geopolitical Eye: Steve Bell and the If… Cartoons," *Political Geography*, 15: 571–92.

_____ (2000), *Geopolitics in a Changing World* (Harlow, England: Prentice Hall/ Pearson Education). Insights into Human Geography, series editors P. Knox and S. Smith.

_____ and Atkinson, D. (eds.) (2000), *Geopolitical Traditions: A Century of Geopolitical Thought* (London and New York: Routledge). Critical Geographies series editors T. Skelton and G. Valentine.

Drummond, P. and Paterson, R. (eds.) (1985), *Television in Transition* (London: British Film Institute).

Duchêne, F. (2001), "Quelle place pour l'Europe dans la politique mondiale?" *Commentaire* 95 (Fall 2001): 603–608.

_____ (2002–2003), "Mars, Vénus et l'Olympe," *Commentaire* 100 (winter): 773–79.

Dumay, J-M. (2004), "Avortement: le planning familial résiste à la « guerre contre les femmes » menée par l'administration Bush," *Le Monde* 14 October 2004, p.4.

Ess, C. (ed.) (1996), *Philosophical Perspectives on Computer-Mediated Communication* (Albany, New York: SUNY Press).

European Commission. (2004), "Key Facts and Figures about the European Union," *Europa* [website] <http://europa.eu.int/comm/publications/booklets/eu_glance/44/index_en.htm> accessed 8/16/2005.

European Union. (2001), "Treaty of Nice: Amending the Treaty on European Union, the Treaties Establishing the European Communities and Certain Related Acts," *Official Journal of the European Communities* March 10, 2001.

Faure, G. (2004a), "Les candidats veulent cacher Michael Moore," *Le Figaro* 27 July 2004, p.3.

_____ (2004b), "La guerre féroce des spots publicitaires," *Le Figaro* 30 September 2004, p.2.

_____ (2004c), "Le comté le plus pauvre du pays vote Bush a 80%," *Le Figaro* 23–24 October 2004, p.2.

_____ (2004d), "Un catholique qui ne séduit pas son Église," *Le Figaro* 29 October 2004, p.4.

_____ and Philippe Gélie. (2004), "Schwarzenegger muscle la convention républicaine," *Le Figaro* 2 September 2004, 5.

Favret-Saada, J. (2005), "Mel Gibson en croisade: comment fabriquer de l'œcumenisme avec une Passion intégriste," *Le Débat* 133 (January–February 2005): 65–81.

Figaro, Le. (2004), "Les Français plébiscitent Kerry," *Le Figaro* 28 October 2004, p.1.

Fiske, J. (1987), *Television Culture* (London: Methuen).

_____ and John Hartley. (1978), *Reading Television* (London and New York: Methuen).

Fontaine, A. (2002–2003), "Ego Nominor Leo," *Commentaire* 100 (winter): 781–84.

Froment-Meurice, H. (2001), "La partie n'est pas finie," *Commentaire* 93 (spring): 13–16.

_____ (2004–2005), "Anniversaires," *Commentaire* 108: 949–55.

Froomkin, A.M. (2003), "Habermas@discourse.net: Toward a Critical Theory of Cyberspace," *Harvard Law Review* 116:3, 751–73.

Fulda, A. (2004), "Chirac ou les avatars de l' « ami américain »," *Le Figaro* 28 October 2004, p.3.

Futuribles. (2002–2004), "La révue Futuribles," *Futuribles* [website] <http://www.futuribles.com/revue.html>

Gallimard. (2005), "Gallimard Publishing Group, English Version," *Gallimard* [website] <http://www.gallimard.fr/catalog/html/PDF/gall-eng.pdf> accessed 10 May 2005.

Gaudemar, A. (2004a), "Second souffle," *Libération* 26 July 2004, p.3.

_____ (2004b), "Se mouiller," *Libération* 30 July 2004, p.3.

Gélie, P. (2004a), "Bush, « président de guerre » en quête d'idées neuves," *Le Figaro* 2 September 2004, p.5.

_____ (2004b), "Bush-Kerry: le choc de deux Amérique," *Le Figaro* 18 October 2004, p.2.

_____ (2004c), "Six enjeux qui divisent les États-Unis," *Le Figaro* 18 October 2004, p.3.

_____ (2004d), "La « rust belt » tente d'échapper au déclin," *Le Figaro* 21 October 2004, p.2.

_____ (2004e), "La bataille à venir pour le contrôle de la justice," *Le Figaro* 23–24 October 2004, p.3.

_____ (2004f), "Le terrorisme au cœur du duel électoral," *Le Figaro* 28 October 2004, p.2.

_____ (2004g), "Bush et Kerry portent leur foi en bandoulière," *Le Figaro* 29 October 2004, p.4.

_____ (2004h), "Deux visions des Etats-Unis," *Le Figaro* 2 November 2004, p.5.

_____ (2004i), "Le duel Bush-Kerry a mobilisé l'Amérique," *Le Figaro* 3 November 2004, p.2.

Giddens, A. (1984), *The Constitution of Society: Outline of the Theory of Structuration* (Berkeley and Los Angeles: University of California Press).

Girard, R. (2004), "Le message et le messager," *Le Figaro* 17 October 2004, p.13.

Gitlin, T., (ed.) (1986), *Watching Television: A Pantheon Guide to Popular Culture* (New York: Pantheon Books).

_____ (1980), *The Whole World Is Watching: Mass Media in the Making and Unmaking of the New Left* (Berkeley, CA, Los Angeles, CA & London, U.K.: University of California Press).

Gordon, P. (2004–2005), "Pour un new deal: lettre à mes amis français sur l'Europe et l'Amérique," *Commentaire* 108 (winter): 915–21.

Gréau, J-L. (2003), "Les contradictions de l'empire," *Le Débat* 123: 32–47.

Grewal, S.S. (2001), "The Paradox of Integration: Habermas and the Unfinished Project of European Union," *Politics* 21:2, 114–23.

Groth, P. and Bressi, T.W. (eds.) (1997), *Understanding Ordinary Landscapes* (New Haven and London: Yale University Press).

Habermas, J. (1984), *The Theory of Communicative Action.* Vol. 1, Reason and the Rationalization of Society. Translated by Thomas McCarthy (Boston: Beacon Press).

_____ (1987), *The Theory of Communicative Action.* Vol. 2, Lifeworld and System: A Critique of Functionalist Reason. Translated by Thomas McCarthy (Boston: Beacon Press).

_____ (1990), "Discourse Ethics: Notes on a Program of Philosophical Justification," in *Moral Consciousness and Communicative Action,* translated by Christian Lenhardt and Shierry Weber Nicholsen, pp.43–115 (Cambridge: MIT Press). Originally "Diskursethik: Notizen zu einem Begruendungsprogram," In *Moralbewusstsein und kommunikatives Handeln* (Frankfurt: Suhrkamp, 1983).

_____ (1991), *The Structural Transformation of the Public Sphere: An Inquiry into a category of Bourgeois Society*, translated by Thomas Burger with Frederick Lawrence (Cambridge, MA: MIT Press).

_____ (1995), "Reconciliation through the Public Use of Reason: Remarks on John Rawls's Political Liberalism," *Journal of Philosophy* XCII:3 (March 1995): 117–8.

_____ (1999), "The European Nation-state and the Pressures of Globalisation," *New Left Review* 235: 46–59. Originally appeared in *Blätter für deutsche und internationale Politik,* April 1999, pp. 425–36.

Hahn, L.E. (ed.) (2000), *Perspectives on Habermas* (Peru, Illinois: Carus/Open Court Press).

Halevi, R. (2004), "France-Amérique: la scène primitive d'une mésintelligence pacifique," *Le Débat* 129 (March/April): 27–48.

Harriss, J.A. (2004–2005), "Bonjour tristesse," *The American Spectator* December 2004–January 2005: 64–65.

Hauter, F. (2004), "Cap sur un rapprochement transatlantique," *Le Figaro* 3 November 2004, p.8.

Heffernan, M. (2000), "Fin de siècle, fin du monde? On the origins of European geopolitics, 1890–1920," *Geopolitical Traditions: A Century of Geopolitical Thought,* K. Dodds and D. Atkinson (ed.), 27–51 (New York and London: Routledge).

Hénaff, M. and Strong, T.B. (2001), "Public Space, Virtual Space, and Democracy," in Hénaff and Strong (eds.).

Hénaff M. and Strong, T.B. (eds.) (2001), *Public Space and Democracy* (Minneapolis: University of Minnesota Press).

Herb, G. and Kaplan, D. (ed.) (1999), *Nested Identities: Nationalism, Territory and Scale* (Lanham, Maryland: Rowman and Littlefield).

Herod, A. et al. (eds.) (1998), *Unruly World: Globalization, Governance and Geography* (New York: Routledge).

Herring, S.C. (1996), "Posting in a Different Voice: Gender and Ethics in Computer-Mediated Communication," in Ess (ed.).

_____ (1999), "The Rhetorical Dynamics of Gender Harassment On-Line," *The Information Society* 15, 151–67.

Hillis, K. (1996), "A Geography of the Eye: The Technologies of Virtual Reality," in Shields (ed.).

Hobbes, T. (1985 [1651]), *Leviathan* (London and New York: Penguin Books).

Hodgson, G. (2005), "La nouvelle Amérique conservatrice," *Le Débat* 133 (January–February 2005): 12–26.

Holland, F. (2004), "Ce modèle conservateur aura des imitateurs," *Le Monde* 5 November 2004, p.12.

Holmes, D. (ed.) (1997), *Virtual Politics: Identity and Community in Cyberspace* (London, Sage).

Horkheimer, M. and Adorno, T.W. (1972), *Dialectic of Enlightenment,* translated by J. Cumming (New York: Continuum). Original publication *Dialektik der Aufklarung,* New York: Social Studies Association, 1944.

Hugo, Victor. 1998 [1837] "À l'Arc de Triomphe," *Les voix intérieures* (web site posted 1998) <http://hypo.ge-dip.etat-ge.ch/athena/hugo/hug_voix.html> accessed 15 August 2006.

Huntington, S.P. (1997), *The Clash of Civilizations and the Remaking of World Order* (New York: Touchstone, Simon & Schuster).

Hureaux, R. (2001), "Les Etats-Unis, l'Allemagne et l'Europe," *Commentaire* 95 (fall): 623–25.

Isin, E.F. and Turner, B.S. (eds.) (2002), *Handbook of Citizenship Studies* (London and Thousand Oaks, California: Sage Publications).

Jarreau, P. (2004a), "Les attentats du 11 septembre 2001 sont au cœur de la campagne présidentielle américaine," *Le Monde* 11 September 2004, p.3.

_____ (2004b), "Le renchérissement de l'assurance-maladie inquiète les Américains," *Le Monde* 13 October 2004, p.5.

_____ (2004c), "L'électorat américain partagé en deux parties égales," *Le Monde* 2 November 2004, p.15.

_____ (2004d), "George W. Bush: l'homme qui ne doute pas," *Le Monde* 3 November 2004, p.18.

_____ (2004e), "John F. Kerry: le battant," *Le Monde* 3 November 2004, p.19.

_____ (2004f), "Avec George Bush réélu, le triomphe de la révolution conservatrice," *Le Monde* 5 November 2004, p.2.

Jaulmes, A. (2004a), "La vague démocrate déferle à Boston," *Le Figaro*, 27 July 2004, p.3.

_____ 2004b. "John Ferry engage le fer contre George Bush," *Le Figaro*, 30 July 2004, p.2.

Johnson, N. (2001), "From Time Immemorial: Narratives of Nationhood and the Making of National Space," in May and Thrift (eds.).

Jonas, A.E.G. (1994), "Editorial: The Scale Politics of Spatiality," *Environment and Planning D: Society and Space* 12:3, 257–64.

_____ (2004), "The scale spatiality of politics," *Environment and Planning D: Society and Space,* 12:3, 257–64.

Jouvenel, H. de. (2004), "Pour une Europe politique," *Futuribles* 303 (Dec 2004): 3–4.

Joxe, A. (2004a), *L'empire du chaos: les Républiques face à la domination américain dans l'après guerre froide* (Paris: Éditions La Découverte/Poche).

_____ (2004b), "L'Empire du chaos est-il militaire?" in Wieviorka (ed.).

Julliard, J. (2005), "Journal des États-Unis: septembre-octobre 2004," *Le Débat* 133 (January–February 2005): 27–45.

Kagan, R. (2002), "Power and Weakness," *Policy Review* 113 (June and July 2002).

_____ (2002), "Puissance et faiblesse," *Commentaire* 99 (fall): 517–35.

_____ (2003), "Napoleonic Fervor," *The Washington Post,* Monday, February 24: A21.

Kant, I. (1983 [1795]), *Perpetual Peace, and Other Essays on Politics, History, and Morals* (Indianapolis, Indiana: Hackett Publishing Company). (HPC Classics Series)

Katz, E. and Dayan, D. (1985), "Electronic Ceremonies: Television Performs a Royal Wedding," in Blonsky (ed.).

Katz, E. and Liebes, T. (1985), "Mutual Aid in the Decoding of Dallas: Preliminary Notes from a Cross-Cultural Study," in Drummond and Paterson (eds.).

Kellner, D. (2000), "Habermas, the Public Sphere, and Democracy: A Critical Intervention," in Hahn (ed.).

Kellner, T. (2003), "What Gaul!" *Forbes* 171, 9 (April 28): 52.

King, A.D. (1997), "The Politics of Vision," in Groth and Bressi (eds.).

Kirby, A. (2000), "The Construction of Geopolitical Images: The World According to Biggles (and other fictional characters)," in Dodds and Atkinson (eds.).

Kissinger, H. (2001), *Does America Need a Foreign Policy? Toward a Diplomacy for the Twenty-First Century* (New York: Simon and Schuster).

Kress, G. (2003), *Literacy in the New Media Age* (New York and London: Routledge).

Kristeva, J. (1980), *Desire in Language: A Semiotic Approach to Literature and Art* (New York: Columbia University Press).

Laclau, E. (1977), *Politics and Ideology in Marxist theory* (London: New Left Books).

Langer, W.R. (1947), *Our Vichy Gamble* (Shoe String Press).

Lanham, R.A. (1993), *The Electronic Word: Democracy, Technology, and the Arts* (Chicago: University of Chicago Press).

Lanxade, J. (2004), "La stratégie américain," in Wieviorka (ed.).

Lazarsfeld, P.F. et al. (1944), *The People's Choice* (New York: Columbia University Press).

Le Guelte, G. (2003), "*Le Monde* de George W. Bush et l'Europe," *Le Débat* 125 (May–Aug 2003): 17–27.

Leicester, J. (2004), "'Punish France'--In Europe, Condoleezza Rice's Reputation Precedes Her," *Associated Press* (16 November 2004), *San Francisco Chronicle* [website] <http://www.sfgate.com> (home page).

Leser, E. (2004a), "Elu, John Kerry aurait à choisir entre le social et la réduction du déficit," *Le Monde* 4 November 2004, p.8.

_____ (2004b), "La priorité accordée aux « valeurs morales » a profité à George Bush," *Le Monde* 5 November 2004, p.5.

Lesnes, C. (2004a), "Le jour le plus long--et, une fois encore, le récit de la demande en mariage de George à Laura," *Le Monde* 3 November 2004, p.3.

_____ (2004b), "La « débâcle » de John Kerry ouvre une ère d'incertitude pour le camp démocrate," *Le Monde* 5 November 2004, p.3.

Libération (2004a), "Chrétien, je vote Bush," *Libération*, 2 November 2004, p.7.

_____ (2004b), "Républicain de base," *Libération*, 2 November 2004, p.7.

_____ (2004c), "Kerry « pour la classe ouvrière »," *Libération*, 2 November 2004, p.9.

_____ (2004d), "Un premier vote contre la crise," *Libération*, 2 November 2004, p.9.

Lieven, A. (2004), "L'Amérique en proie au nationalisme," *Le Débat* 130 (May–Aug): 21–30.

_____ (2005), "Les composantes du nationalisme américain," *Le Débat* 133 (January–February 2005): 46–52.

Lord, C. (1998), D*emocracy in the European Union* (Sheffield: Sheffield Academic Press).

_____ (2001), "Assessing Democracy in a Contested Polity," *Journal of Common Market Studies* 39:4, 641–61.

Low, M. (1997), "Representation Unbound: Globalization and Democracy," in Cox (ed.).

Luce, H. (1941), "The American Century," *Life,* 17 February: 61–65.

Luke, T and Ó Tuathail, G.. (1998), "Global Flowmations, Local Fundamentalisms, and Fast Geopolitics: 'America' in an accelerating world order," *An Unruly World: Globalization, Governance and Geography,* A. Herod, G. Ó Tuathail and S. M. Roberts (eds.), 72–94 (London and New York: Routledge).

Magnette, P. (2003), "European Governance and Civic Participation: Beyond Elitist Citizenship?" *Political Studies* 51: 144–160.

Maïla, J. (2003), "Le cours de la guerre," *Etudes* (May): 581–84.

Majone, G. (1998), "Europe's 'Democratic Deficit': The Question of Standards," *European Law Journal* 4:1, 5–28.

Marti, S. (2004), "L'Amérique vote aussi sur l'économie," *Le Monde* 3 November 2004, p.Economie I.
May, J. and Thrift, N. (eds.) (2001), *Timespace: Geographies of Temporality* (London and New York: Routledge).
McLuhan, M. (1962), *Gutenberg Galaxy* (Toronto: University of Toronto Press).
Meadowcroft, J. (2002), "The European Democratic Deficit, the Market and the Public Space: A Classical Liberal Critique," *Innovation: The European Journal of Social Sciences.* 15:3, 181–192.
Melandri, P. (2003), "L'unilateralisme, stade suprême de l'exceptionnalisme?" *Le Débat* 127 (Nov–Dec): 21–45.
Michelot, V. (2005), "Élections 2004: Un plébiscite en trompe l'œil ?" *Le Débat* 133 (January–February 2005): 53–64.
Miller, J.J. and Molesky, M. (2004), *Our Oldest Enemy: A History of America's Disastrous Relationship with France* (New York: Doubleday).
Mitchell, D. (1996), *The Lie of the Land: Migrant Workers and the California Landscape* (Minneapolis: University of Minnesota Press).
Mitchell, W.J. (1995), *City of Bits: Space, Place and the Infobahn* (Cambridge, Massachusetts: MIT Press).
Moïsi, D. (1995), "L'Amérique dans les relations franco-allemandes," in de Montbrial et al. (eds.).
Monar, J. (2004), "The EU as an International Actor in the domain of Justice and Home Affairs," *European Foreign Affairs Review* 9: 395–415.
Monde, Le. (2004), "Une enquête conduite par dix journaux," *Le Monde* 16 October 2004, p.18.
Monde, Le. (2004a), "Des extrémistes chrétiens en campagne pour Bush," *Le Monde* 29–30 August 2004, p.1.
Monde, Le. (2004b), "Le choix américain," *Le Monde* 31 October–1 November 2004, p.15.
Monroe, J. (1823), "President Monroe's Seventh Annual Message to Congress," (December 2, 1823) The University of Oklahoma Law Center, College of Law, The University of Oklahoma <http://www.law.ou.edu/hist/monrodoc.html> accessed 15 August 2006.
Moravcsik, A. (2004), "Is there a 'Democratic Deficit' in World Politics? A Framework for Analysis," *Government and Opposition* 39:2, 336–63.
Newman, D. (2000), "Citizenship, Identity and Location: The Changing Discourse of Israeli Geopolitics," in Dodds and Atkinson (eds.).
Nora, P. (2000), "Vingt ans de Débat," *Le Débat* <http://www.le-debat.gallimard.fr> (home page) accessed 22 November 2005.
Nowina-Konopka, P. (2003), "Democratic Deficit: A European Scapegoat for Domestic Trouble," *European Foreign Affairs Review* 8: 1–4.
Nye, J.S. (2002), *The Paradox of American Power: Why the World's Only Superpower Can't Go It Alone* (Oxford, UK: Oxford University Press).
_____ (2004), *Soft Power: The Means to Success in World Politics* (New York: Public Affairs).

Ó Tuathail, G. (1992), "Foreign Policy and the Hyperreal: The Reagan administration and the scripting of 'South Africa'," in Barnes and Duncan (eds.).

_____ (1996), *Critical Geopolitics: the politics of writing global space* (Minneapolis: University of Minnesota Press).

_____ (2000), "Spiritual Geopolitics: Fr. Edmund Walsh and Jesuit Anti-Communism," in Dodds and Atkinson (eds.).

_____ and Agnew, J., (1998), "Geopolitics and Discourse: Practical Geopolitical Reasoning in American foreign policy," in Ó Tuathail et al. (eds.).

_____ et al., (eds.) (1998), *The Geopolitics Reader* (New York: Routledge).

_____ and Dalby, S. (1998), "Introduction: Rethinking Geopolitics: Towards a Critical Geopolitics," in Ó Tuathail and Dalby (eds.).

_____ and Dalby, S. (eds.) (1998), *Rethinking Geopolitics* (London and New York: Routledge).

_____ and Luke, T.. (1994), "Present at the (Dis)Integration: Deterritorialisation and Reterritorialisation in the New Wor(l)d Order," *Annals of the Association of American Geographers* 84: 381–94.

O'Lear, S. (1997), "Electronic Communication and Environmental Policy in Russia and Estonia," *The Geographical Review* 87:2, 275–290.

Olwig, K. (2001), "Landscape as a Contested Topos of Place, Community and Self," in Adams et al. (eds.).

Parker, G. (1987), "French Geopolitical Thought in the Interwar Years and the Emergence of the European Idea," *Political Geography Quarterly,* 6:2, 145–50.

Parmentier, G. (2002–2003), "Force, faiblesse, puissance ?" *Commentaire* 100 (winter): 791–99.

Paye, J-C. (2001), "Etats-Unis et Europe à l'OCDE," *Commentaire* 94 (summer): 313–14.

Pew (The Pew Research Center for People and the Press) (2003), "America's Image Further Erodes: Europeans Want Weaker Ties," survey report (18 March 2003) <http://people-press.org/reports/display.php3?ReportID=175> accessed 15 August 2006.

Poniatowski, A. (2004), *Pourquoi les Francais et les Américains ne se comprennent plus* (Paris: Editions Perrin).

Postel-Vinay, K. (2005), "Récits géopolitiques pour le XXIe siècle," *Etudes* 4021 (January): 19–26.

Poster, M. (1997), "Cyberdemocracy: Internet and the Public Sphere," In Holmes (ed.).

_____ (2001), *What's the Matter with the Internet?* (Minneapolis: University of Minnesota Press). Electronic Mediations Volume 3.

Postman, N. (1985), *Amusing Ourselves to Death: Public Discourse in the Age of Show Business* (New York: Penguin Books).

Raffarin, J-P. (2005), Interview on the EuroNews Channel <http://www.premier-ministre.gouv.fr/en/acteurs/speeches_45/an_interview_with_prime_52340.html> accessed 15 August 2005.

Raynaud, P. (2001), "États-Unis et Europe: limites d'une comparaison," *Commentaire* 95(fall): 613–616.

Relph, E. (1976), *Place and Placelessness* (London: Pion, Ltd.).

Revel, J-F. (2002), *L'obsession anti-américain: son fonctionnement, ses causes, ses inconséquences* (Paris: Plon).

Rheingold, H. (1993), *The Virtual Community: Homesteading on the Electronic Frontier.* (Reading, Massachusetts: Addison Wesley), "A William Patrick Book,"

Richardson, J.L. (1992), *The Gulf War and Australian Political Culture.* Research School of Pacific Studies *(*Australian National University, Canberra).

Riché, P. (2004a), "Bush s'empêtre dans son mensonge," *Libération* 8 October 2004, p.6.

_____ (2004b), "Bush mise sur les slogans, Kerry tente les arguments," *Libération* 11 October 2004, p.10.

_____ (2004c), "Deux Amériques dans un mouchoir de poche," *Libération* 25 October 2004, p.2–3.

_____ (2004d), "Bush et Kerry rivalisent d'économies," *Libération* 29 October 2004, p.10–11.

_____ (2004e), "Bush, Dieu et la droite dure pour credo: Obnubilé par la faiblesse de son père et obsédé par sa foi, le président a radicalisé son discours," *Libération* 2 November 2004, p.6.

_____ and Rousselot, F.. (2004), "Bush reprend l'étendard du 11 septembre," *Libération* 31 August 2004, p.6.

Riotta, G. (2000), "The Coming Identity War," *Foreign Policy*, September–October 2000: 87.

Robin, C. (2004), "L'administration Bush gouverne par la peur, qui est devenue le seul langage de la vie publique," *Le Monde* 12–13 September 2004, p.4.

Roger, P. (1996), *Rêves et cauchemars américains: les États-Unis au miroir de l'opinion publique française (1945–1953),* (Villeneuve d'Ascq, France: Presses Universitaires du Septentrion).

_____ (2002), *L'Ennemi américain: généalogie de l'antiaméricanisme français* (Paris: Éditions du Seuil).

_____ (2004), "« Empire américain », fantasmes français," in Wieviorka (ed.).

Roper (The Roper Center for Public Opinion Research), (2005), "News Interest Index Poll," (October 2005) <http://www.ropercenter.uconn.edu> accessed October 2005.

Rousselot, F. (2004), "John Kerry l'insaisissable: distant, le candidat démocrate est en proie à des contradictions," *Libération* 26 July 2004, p.3–4.

Rubenfeld, J. (2004), "Deux visions de l'ordre mondial," *Le Débat* 129 (March–April): 17–26.

Rumford, C. (2001), "Social Spaces beyond Civil Society: European Integration, Globalization and the Sociology of European Society," *Innovation* 14:3, 205–18.

Sabatier, P. (2004), "Rouges et bleus," *Libération* 16–17 October 2004, p.5.

Said, E. (1994), *Orientalism* (New York: Vintage Books).

Saint-Étienne, C. (2003), *La puissance ou la mort: l'Europe face à l'empire Américain* (Paris: Éditions du Seuil).

Schlesinger, P. (1999), "Changing Spaces of Political Communication: The Case of the European Union," *Political Communication* 16: 263–79.

Schuler, D. (ed.) (2003), *Digital Directions* (Cambridge, MA: MIT Press).

Sclove, R.E. (1995), *Democracy and Technology* (New York and London: The Guilford Press).

Sequence France. (2004), "Le « tout sauf Bush » domine à gauche comme à droite," *Le Monde* 2 November 2004, p.7.

Shapiro, M.J. (1999), *Cinematic Political Thought: Narrating Race, Nation and Gender* (New York: New York University Press).

Sharp, J. (2000), *Condensing the Cold War: Reader's Digest and American Identity* (Minneapolis: University of Minnesota Press).

Shields, R. (ed.) (1996), *Cultures of Internet: Virtual Spaces, Real Histories, Living Bodies* (Thousand Oaks, California: Sage Publications).

Sidaway, J.D. (2000), "Iberian geopolitics," in Dodds and Atkinson (eds.).

Smith, R.M. (2002), "Modern Citizenship," in Isin and Turner (eds.).

Smith, M.A. and Kollock, P. (eds.) (1999), *Communities in Cyberspace* (London and New York: Routledge).

Sontag, S. (1977), *On Photography* (New York: A Delta Book/Farrar, Straus & Giroux, Inc.).

Sorman, G. (2004), "Dieu est-il américain ?" *Le Figaro* 1 October 2004, p.12.

Soulé, V. (2004), "La France s'accommoderait de l'incommodant Bush," *Libération* 3 November 2004, p.3.

Staeheli, L. (1994), "Empowering Political Struggle: Spaces and Scales of Resistance," *Political Geography* 13: 387–91.

Stahl, B. et al. (2004), "Understanding the Atlanticist-Europeanist Devide in the CFSP: Comparing Denmark, France, Germany and the Netherlands," *European Foreign Affairs Review* 9: 417–441.

Starrs, P.F. (1997), "The Sacred, the Regional, and the Digital," *The Geographical Review* 87:2, 193–218.

Stern, B. (2004), "Près de 36 millions d'Américains vivent sous le seuil de pauvreté," *Le Monde* 28 August 2004, p.3.

Takeuchi, K. (2000), "Japanese geopolitics in the 1930s and 1940s," in Dodds and Atkinson (eds.).

Tatum, S. (1982), *Inventing Billy the Kid: Visions of the Outlaw in America, 1881–1981* (Albuquerque: University of New Mexico Press).

Taylor, P. (1995), "Beyond Containers: Internationality, Interstateness, Interterritoriality," *Progress in Human Geography* 19:1, 1–15.

Thomas, E. (2006), "Keeping Identity at a Distance: Explaining France's New Legal Restrictions on the Islamic Headscarf" *Ethnic and Racial Studies* 29:2, 237–59.

Timmerman, K. (2004), *The French Betrayal of America.* (New York: Crown Forum).

Tincq, H. (2004), "Protestants évangéliques et catholiques conservateurs unis dans le vote gagnant," *Le Monde* 5 November 2004, p.6.

Tocqueville, A. de. (1969) [1835–1840], *Democracy in America*, translated by G. Lawrence, edited by J.P. Mayer. (New York: HarperCollins/Perennial Classics) (Original edition, Paris: Gallimard, Vol. 1, 1835; Vol. 2, 1840).

Todd, E. (1976), *La chute finale: essai sur la décomposition de la sphère soviétique* (Paris: R. Laffont).

_____ (1998), *L'Illusion économique: essai sur la stagnation des sociétés dévelopées* (Paris: Éditions Gallimard).

_____ (2002), *Après l'empire: essai sur la décomposition du système américain* (Paris: Éditions Gallimard).

_____ (2003), "Les États-Unis, l'Europe et le capitalisme," *Le Débat* 123 (January–February): 66–69.

Toulemon, R. (2005), "De la construction européen à la réforme de Nations Unies," in Albert (ed.).

Tréan, C. (2004a), "Une forte majorité attend de John Kerry un tournant en Irak," *Le Monde* 16 October 2004, p.18.

_____ (2004b), "L'opinion mondiale souhaite, en majorité, la défaite de George Bush," *Le Monde* 3 November 2004, p.8.

Tuan, Y-F. (1974), *Topophilia,* Morningside Edition (New York: Columbia University Press).

Turkle, S. (1995), *Life on the Screen: Identity in the Age of the Internet* (New York: Simon and Schuster).

Van Apeldoorn, B. (2003), "The Struggle over European Order: Transnational Class Agency in the Making of 'Embedded Neo-Liberalism'," in Brenner, et al. (eds.).

Van der Wusten, H. andDijkink, G. (2002), "German, British and French Geoplitics: The Enduring Differences," *Geopolitics* 7:3, 19–38.

Védrine, H. (2003a), "Le défi du déclin?" *Le Débat* 123 (Jan–Feb 2003): 63–5.

_____ (2003b), "Que faire avec l'hyperpuissance?" (interview) *Le Débat* 125 (May–Aug 2003): 4–16.

_____ (2003c), *Face à l'hyperpuissance: textes et discours 1995–2003* (Paris: Fayard.

_____ (2004), "On Anti-Americanism," *Brown Journal of World Affairs* 10:2, (Winter/Spring): 117–21.

Vernet, D. (2002–2003), "Le dilemma européen," *Commentaire* 100 (Winter 2002–2003): 801–805.

Villepin, D. de. (2004), *Le Requin et la Mouette* (Paris: Plon/Albin Michel).

Vincent-Arriola, A. (2004), "Les homosexuels divisés face aux options morales du président sortant," *Le Figaro* 29 October 2004, p.4.

Virilio, P. (1989), *War and Cinema.* Translated by Patrick Camiller (London and New York: Verso).

Warf, B. and Grimes, J. (1997), "Counterhegemonic discourses and the Internet," *Geographical Review* 87:2, 259–74.

Wellman, B. and Gulia, M.. (1999), "Virtual communities as communities," in Smith and Kollock (eds.).

Westley, B.H. and MacLean, M. (1957), "A Conceptual Model for Mass Communication Research," *Journalism Quarterly* 34: 31–38.

Wieviorka, M. (ed.) (2004), *L'Empire américain?* (Paris: Éditions Balland).

Williams, R. (1974), *Television: Technology and Cultural Form* (New York: Schocken Books).

Winchester, H. (1993), *Contemporary France* (Harlow, UK: Longman).

Wood, Allen W. (1998), "Kant's Project for Perpetual Peace," in Cheah and Robbins (eds.).

Young, I.M. (1986), "The Ideal of Community and the Politics of Difference," *Social Theory and Practice,* 12:1 (Spring), 1–26; reprinted in L. Nicholson, (ed.), *Feminism/Postmoderism* (New York, Routledge, Chapman and Hall, 1990) 300–326; reprinted in P.A. Weiss and M. Friedman, (eds.), *Feminism and Community* (Philadelphia: Temple University Press, 1995).

Zorgbibe, C. (2003), "Pourquoi l'OTAN ?" *Etudes* (April): 451–64.

Index

For Product Safety Concerns and Information please contact our EU representative GPSR@taylorandfrancis.com
Taylor & Francis Verlag GmbH, Kaufingerstraße 24, 80331 München, Germany

www.ingramcontent.com/pod-product-compliance
Lightning Source LLC
LaVergne TN
LVHW011125300926
849160LV00008B/513

* 9 7 8 1 1 3 8 6 1 8 9 0 9 *